A Chanticleer Press Edition

Edward R. Ricciuti

WILDLIFE OF THE MOUNTAINS

HARRY N. ABRAMS, INC., PUBLISHERS, NEW YORK

1-9. *The mountains of the world, especially in the heights, are among the last strongholds of primal wilderness. Although even the mountains are not unassailed by environmental changes brought by man, the peaks are still a world where nature remains unspoiled. In the heights, one can still find scenes of primitive beauty, as primal as before the advent of man. Creatures such as the Rocky Mountain goat* (Oreamnos americanus) *and mouflon* (Ovis ammon musimon) *inhabit a realm that, while breathtaking in its splendor, is nevertheless hostile. Within the mountain realm, birds such as the Andean condor* (Vultur gryphus) *soar overhead and hunting beasts such as the cougar* (Felis concolor) *and prey such as the bighorn sheep* (Ovis canadensis) *go about the endless struggle for existence just as they have for many thousands of years.*

Library of Congress Catalogue Card Number: 78–19873
Ricciuti, Edward R.
Wildlife of the Mountains.
A Chanticleer Press Edition.
Includes index.
1. Mountain ecology. I. Title.
QH541.5.M65R53   574.5'264   78–19873
ISBN 0–8109–1757–2

Prepared and produced by Chanticleer Press, Inc., New York
Color reproductions by Fontana & Bonomi, Milan, Italy
Printed and bound by Amilcare Pizzi, S.p.A., Milan, Italy
Composition by American–Stratford Graphic Services, Inc., Brattleboro, Vermont

Note: Illustrations are numbered according to the pages on which they appear.

Staff of this book:
Prepared and Produced by Chanticleer Press:
Publisher: Paul Steiner
Editor-in-Chief: Milton Rugoff
Managing Editor: Gudrun Buettner
Project Editor: Mary Suffudy
Production: Helga Lose, Ray Patient
Art Associates: Carol Nehring, Johann Wechter
Maps: Richard Edes Harrison, Courtesy of Nystrom, Division Carnation Co.
Drawings: Howard Friedman, Paul Singer
Design: Massimo Vignelli

Contents

Introduction: There Shall Be Mountains 16

Life in Layers 22
Life in the High Places of Europe 40
The Heights of Africa 96
Asia and the South Pacific 118
The Americas:
The Great Cordillera and Other Ranges 148

Glossary 194
High Places of the World 214
Index 224
Photographic Credits 231

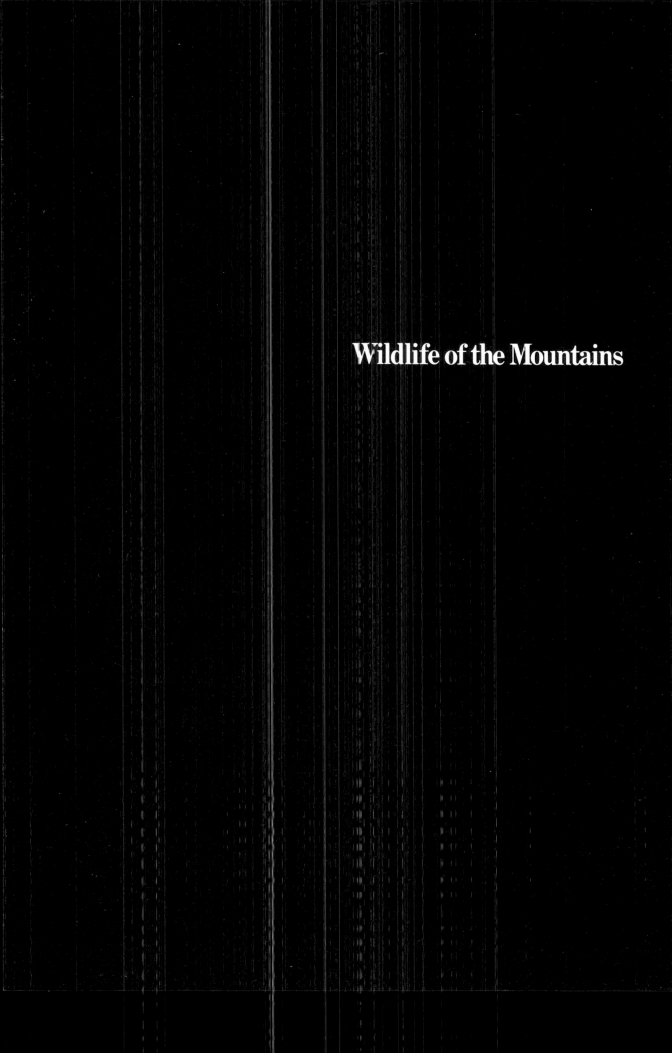

Wildlife of the Mountains

The Earth has a solid inner core, liquid outer core, and a dense but plastic mantle covered by a thin crust. Convection currents caused by differences in temperature within the mantle are believed to cause crustal deformation. Where convection currents well up, the crust is lifted up, resulting in such formations as the great mid-ocean ridges.

Where the currents flow downward, deep ocean trenches are formed. New oceanic crust is added to the ridges and pulled down into the trenches. These shifts in the crust help account for the movements now known as continental drift.

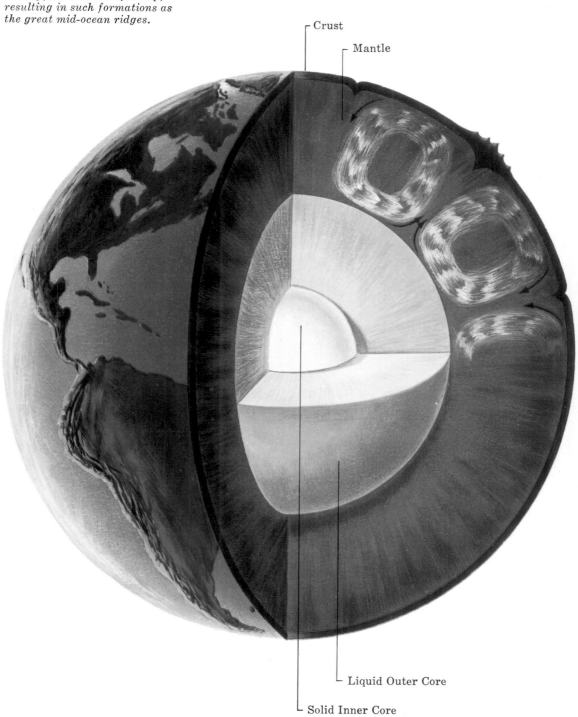

Crust

Mantle

Liquid Outer Core

Solid Inner Core

Introduction:
There Shall Be Mountains

Floating in the emptiness beyond a gray precipice, an alpine chough (*Pyrrhocorax graculus*), black feathers ruffled, rides the currents of the mountain winds, and soars and loops without apparent effort. Moving with incredible fluidity, a Rocky Mountain goat (*Oreamnos americanus*) ascends a nearly vertical wall of rock on the roof of western North America. Atop a pinnacle, the goat pauses, its shaggy coat as white as the glaciers that cap the heights. It is a monarch of the high places.

Such sights, of alpine animals in their sky-high habitats, evoke envy and awe in us—envy because we will never attain such freedom in the heights, and awe because the vastness of scale is beyond anything we encounter in our daily lives. Everyone who has ever been high on a mountain knows the exhilaration that such heights inspire, and the wonder that nature has produced such colossal landforms.

Forces Beneath the Crust

Enormous amounts of energy were needed to rearrange the Earth into mountains. Although scientists have been studying mountains for centuries, the forces that power the processes of mountain building remain rather mysterious. It is known, however, that these processes are ultimately the result of events occurring deep below the crust of the Earth.

The crust of the Earth can be compared to the skin of an orange. While the diameter of the Earth is almost 12,900 kilometers, the crust is less than 50 kilometers thick under the floor of the sea. Below the crust is the mantle, a region of very dense rock that is quite solid, but is nevertheless believed to move, although ever so slowly.

Within the mantle, the rock is subjected to tremendous heat and pressure. Geophysicists believe that the heat, generated by radioactive elements in the Earth, creates what are known as convection currents in the mantle rock. The currents are the result of differences in temperature within the mantle. The upper levels of the mantle are much cooler than the lower regions, which are nearer the sizzling core of the Earth. Just as water rises from the bottom of a pot that has been heated, rock from the mantle rises toward the upper levels, just below the crust. The current takes the rock parallel to the crust until cooling sets in, and the rock sinks to the depths, where it is heated, and rises once more, resuming the cycle.

Geophysicists theorize that there are many such currents within the mantle, each forming its own convection cell. Where the sides of the cells come into contact, impact is exerted upon the crust above. If the currents of adjoining cells flow downward, the overlying crust is squeezed. Where the currents travel upward, the crust is spread.

The convection current concept has been used to explain the theory of continental drift, proposed in 1912 and refined over the years until it now has gained great acceptance among scientists. Basically, the theory states that the continents sit on great crustal plates, lower in density than the plastic mantle, and that they ride about on the convection currents flowing parallel to the crust.

Although no theory of the forces behind mountain building has been really proven, there is growing agreement that convection currents and continental drift are involved. The stirrings of the Earth beneath the crust and the movement

1.

2.

3.

4.

5.

1. *Where convection currents well up, a rift may form and the plates composing the Earth's continental crust spread. At the same time molten rock flowing up through the rift adds to the oceanic crust. The plates spread apart at this "divergent boundary."*

2. *Where convection currents flow downward, continental crustal plates may fall into oceanic trenches that have been opened up by the currents. This plate has fallen into a trench near a chain of oceanic islands. The friction caused by the crustal movements creates earthquakes and vulcanism.*

3. *When the oceanic crust plunges below a continental plate, the downward-flowing currents sometimes cause another type of "convergent boundary." The collision of two pieces of crust can uplift mountains at the edge of a continental plate as the rock of the sinking plate is heated and extruded upwards by vulcanism. The Andes were formed in this way.*

4. *Continental plates can sometimes push up mountains when they collide. The Himalayas were formed when the Indian plate rammed into the Asian plate, and the edges of both plates crumpled upwards.*

5. *Colliding continental plates sometimes make new continental crust by pushing up the sea floor.*

of the great plates must, scientists generally believe, trigger the processes that give birth to mountains.

Mountain Building Processes

There are three basic types of mountain building processes —folding, faulting, and vulcanism. Folding and faulting have been strongly linked to the movement of the crustal plates, which also probably contributes to the pressures that force molten rock through the crust and thus create volcanoes.

Folding, most clearly linked to continental drift, is an upward warping of the crust that can be likened to the way a tablecloth wrinkles when pushed across a tabletop. This process tends to build along ranges of mountains. The basic architecture of the Alps, Himalayas, Rockies, Appalachians, Atlas, and Jura mountains, to name a few, is a product of folding. The Jura, part of a larger chain including the Vosges and Black Forest mountains, represent folding at its simplest. When rocks are exposed in the Jura, rows of parallel folds like a succession of waves can be clearly seen.

It is believed that folding takes place when the edges of drifting plates collide and one overrides the other. The edge of the overriding plate, according to this line of thinking, crumples up into fold mountains. The Alps are now considered to have been formed by several collisions between the plates carrying Africa and Europe. The Himalayas seem to have resulted from a collision between the Indian and Asian plates.

In the early 1970s some geologists suggested that the movement of plates creates mountains in another way as well. When plates move away from one another, great rifts, or faults, are formed. Examples of these are the mid-ocean trenches and the Great Rift Valley of Africa. Scientists have suggested that when a trench opens in the ocean crust near the edge of a continent, the gap may swallow a portion of the continental plate, while heated rock from the mantle rises over the edge of the plate to form mountains. The Andes may have been built in this way.

The rending of the Earth's crust produces faulted mountains, which arise when there is a slippage between two masses of rock along a fault, or crack in the Earth. The rocks, however, can move in many ways with respect to each other. Along the Rhine River, in the Great Rift Valley, and in the Sierra Nevada of western North America, fault blocks have been raised into imposing mountain ranges. Some, such as the Black Forest Mountains overlooking the down-faulted block, or graben, of the Rhine, and the Sierra Nevada, have been tilted steeply as well as faulted. Such mountains, distinguished by the name "fault-scarp," slope gently up from the side that is tilted into the Earth, and end abruptly at a cliff above the graben. Faulting often has accompanied folding, or has further sculpted the face of folded ranges after their formation. The Alps and Rockies, for instance, show the effects of faulting, but originated through folding.

Volcanoes

About 10,000 of the world's mountains are volcanoes, 600 of which are active. Volcanoes are formed by the extrusion of molten rock, which beneath the surface is called magma, but when it breaks through the crust and loses most of its volatile gases it is called lava. As the lava bursts from the surface, it is accompanied by steam and immense quantities of ashes and cinders, often flung skyward by monumental explosions. When Krakatoa, an island between Java and Sumatra, erupted in August 1883, the explosions were heard more than 4,800 kilometers away, and steam and ash spread over a region of almost 5,000 square kilometers. When the eruption subsided, the island had been replaced by a hole in the sea bottom, fringed by three tiny islets.

Some volcanic eruptions are relatively quiet, with the lava usually spilling out of fissures and spreading and cooling in thin sheets, one over the other. Sheet after sheet builds up a type of volcano without a sharp summit, with very gentle slopes and a broad base. This sort of volcano, typified by Hawaii's Mauna Loa, is called a "shield volcano," because of its resemblance to a rounded medieval shield. Volcanoes such as Mt. Vesuvius in Italy, Fujisan in Japan, and Mt. Hood in the United States are of another type, called the "composite volcano," because they are formed by both the internal seepage of lava from channels running out like spokes from the central cone, and the explosion of lava, cinders, and ash from the main crater.

Sometimes lava solidifies in the conduits of volcanic cones, plugging them. Erosion may eventually strip away the surrounding volcano, leaving the plugs, or necks, exposed as mountains in their own right. A classic example of this sort of landform is Ship Rock, in the state of New Mexico. Lava also sometimes builds up just beneath the surface, pushing up the thin skin above it into a giant blister, which eventually wears away, leaving a great dome of cooled lava.

World Without Mountains

To humans, the immense, towering bulk of great mountains may seem immutable, but this is a deception. While it is true that from our perspective, vast ranges such as the Alps and the Himalayas endure almost beyond imagining, in terms of geologic time mountains are as transitory as other forms of nature. Many times in the history of the Earth, mountains have reared their stony summits at least as high as any existing peaks and then have subsided, to be lost in the basements of the continents or under the floor of the sea.

It is difficult to imagine a world without mountains, but until the crust of the Earth solidified, perhaps more than 4 billion years ago, no mountains existed. The first mountains were formed of the rock that now underlies the continents and forms great platforms called "shields." The Canadian shield, for instance, reaches from the states of Minnesota and Wisconsin in the United States, around Hudson Bay to Labrador and Greenland, and touches the northern tip of Scotland. Much of this ancient rock—called "Precambrian" because it was formed before the beginning of the Cambrian period, 600 million years ago—is buried deep within the Earth. With the shield regions, however, there are great expanses of Precambrian outcroppings. Precambrian rock also can be seen where the earth has been slashed by great gorges, such as the Grand Canyon of Arizona in the United States. Starting at slightly more than 600 meters down the wall of the Canyon, a mass of Precambrian schist rock continues all the way to the bottom,

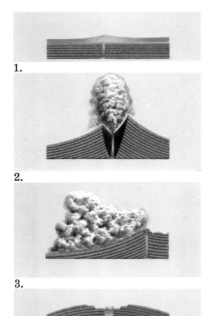

1.

2.

3.

4.

There are several forms of vulcanism.
1. The Icelandic (or Fissure) type occurs when lava seeps through a crack in the crust and spreads out in thin sheets.
2. Strombolian vulcanism occurs when lava fountains form a crater.
3. Composite (or Vulcanian) vulcanism is marked by violent explosions of lava from the central crater and extrusions of lava from channels flowing out of the central cone.
4. Shield (or Hawaiian) volcanoes build up from layers of lava that have gradually seeped from craters.

Kilauea, top, *on the island of Hawaii, and Nyragongo,* bottom, *in Zaïre, are among the world's 600 active volcanoes.*

where it forms the bed of the Colorado River. Much of the Precambrian rock is schist, a rock containing large amounts of mica, quartz, and hornblende. Schist is arranged in leaf-like layers which are prone to split, and is a name taken from a Greek term meaning "easily cleft."

From studying exposed Precambrian rocks, geologists have discovered that between 2 billion and 700 million years ago, great mountain ranges were raised and then destroyed half a dozen times in North America and northern Europe alone. Today, the only evidence of these towering mountain chains, their upheavals and subsidences, are convolutions and breaks in layers of rock.

The Ancient Mountains

The oldest mountains that still can be called such are in Scandinavia and the British Isles, and rose more than 330 million years ago. These mountains, which include the Scottish Highlands and the spine of uplands between Sweden and Norway, are low compared with ranges such as the Alps, and generally rise no more than about 1,000 meters, although some reach an altitude of more than 2,000 meters. When these mountains first arose (330 million years ago), they were part of a soaring alpine wall, which today is known as the Caledonian Range, that encircled the North Atlantic Ocean, from eastern North America through Greenland, to Europe. The Caledonian Range was created by folding and faulting which took place during a general period of upheaval. At that time great faults sundered the crust, forming the deep gorges that cradle the lochs of Scotland and the fjords of Norway.

Mountain Arcs and Chains

Today, most of the world's major mountain ranges form arcs, which are strung out one after the other as links in even larger mountain chains. The Himalayas, for instance, are an arc at the center of the main Eurasian chain. It begins in the East Indies, sweeping up through Sumatra to Indochina and Burma, then through the Himalayas, the Zagros Mountains of Persia, the Taurus Mountains of Turkey, and the Balkans into central Europe, where it meets the Alps.

The greatest of all mountain belts is the one that circumscribes the Pacific Ocean and is often called the "ring of fire" because it is aquiver with volcanoes and earthquakes. Not all of these mountains, which include ranges such as the Cascades in the United States and the Andes in South America, are based on continents. Most of the peaks in the western portion of the circum-Pacific belt rise from the sea floor. Only the tallest peaks of these undersea arcs protrude above the waves. Most of these peaks are volcanic and form islands such as the Aleutians and the Marianas.

The Hawaiian Islands are a typical example of islands that really are the summits of colossal undersea volcanoes. The main islands of Hawaii are dominated by the peaks of Mauna Loa and Mauna Kea, looming more than 4,000 meters above the landscape. The altitude above sea level of these peaks makes them giants, but their true immensity becomes apparent only when one considers that their bases are another 4,800 meters below the surface of the sea. If Mauna Loa and Mauna Kea were based on dry land they would be at least as high as Mt. Everest, and probably higher. Altitude, however, does not always determine whether terrain can be described as mountainous, although technically only land above 1,000 meters in altitude can be described as highlands. According to this definition, more than a quarter of the land surface of the Earth consists of highlands, but these are not always mountainous. For instance there are vast plains in western North America and central Asia well above the 1,000-meter level, but not mountainous by any stretch of the imagination; in fact, their surfaces remain relatively flat throughout. On the other hand in many places—some of the West Indies for example—sheer-sided prominences tower above the surrounding landscape, although really not high enough to qualify as highlands.

The relative height, called relief, as well as altitude determines whether or not land can be considered mountainous. So, moreover, does human perspective. To a lowlander, for instance, even a small hill can be a mountain, but to a resident of a truly alpine region a mountain is something rising thousands of meters above the rest of the world, topped by crags and glaciers.

Mountains around the world range in appearance from worn gentle mounds like those of the Alleghenies in the northeastern United States and the Northern Range in Trinidad to the jagged steep heights of the Himalayas and the immense precipices which tower over the lowlands of Natal in South Africa. This great variation among the mountains of our planet is perhaps the most dramatic evidence that, although mighty, mountains do not endure forever.

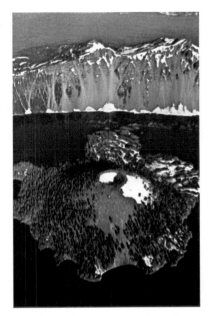

The Cascades of the northwestern United States are a young mountain range characterized by volcanoes, such as the one in the lake in the foreground, and rough, jagged peaks such as those in the background.

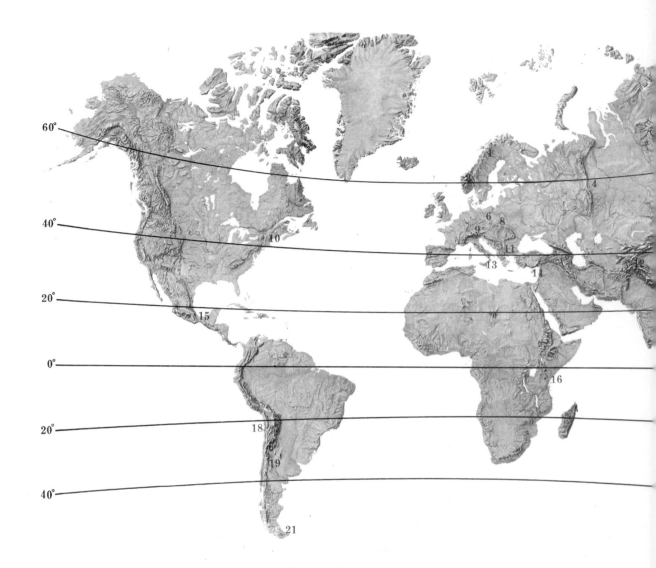

Mountain/Range	Upper Limit of Trees

The broad influences that determine the position of timberlines around the world are altitude and latitude. Many other factors, however, affect how high on mountain slopes trees are found. Timberline tends to be higher, for example, on sunny southerly slopes. Even on the same mountain, timberline will be lower on a slope exposed to strong winds than on the side of the peak in the lee. Slopes prone to avalanches have lower timberlines than those on which avalanches are infrequent. Loose soil—such as volcanic rubble—also reduces the altitude at which trees can gain a foothold. Representative ranges and peaks have been spotted on the map to show gradual elevation in timberline as latitude declines. Approximate upper limits of trees are given in meters.

Mountain/Range	Upper Limit of Trees
1. Sarek Mountains (Europe)	750
2. Jötun-Fjeld (Europe)	900
3. Kamchatka (Asia)	300
4. Ural Mountains (Europe)	1100
5. Sayan Mountains (Asia)	2200
6. Sudetes (Europe)	1250
7. Altai Mountains (Asia)	1100–2400
8. Tatra (Europe)	1600
9. Alps (Europe)	2050
10. Mt. Washington (North America)	1290
11. Balkan Mountains (Europe)	1850
12. Pamirs (Asia)	3600
13. Mt. Etna (Europe)	2200
14. Lebanon Mountains (Asia)	1950
15. Pic-Orizaba (North America)	3200
16. Kilimanjaro (Africa)	3000
17. Java (Asia)	2800
18. Andes Mountains (South America)	2400
19. Aconcagua (South America)	3500
20. Southern Alps (New Zealand)	1300
21. Tierra del Fuego (South America)	550

Life in Layers

Each mountain, or range, has its own special personality. Many factors shape the character of particular mountains, but perhaps nothing influences their aspect more profoundly than their geological history. Low, old mountains such as the Jura and Appalachians are green to the top, while high, young ones such as the Alps and Rockies are crested by frozen pinnacles where nothing at all can live. Clearly, the geology of a mountain helps determine what lives on it. In turn, the plants with which the slopes and scarps of a mountain are clad, and the animals living on them, add to its distinctive qualities.

Whether mountains are young or old, high or low, each is banded from bottom to top by a series of well-defined life zones, with each zone distinguished by its own particular association of plants. This characteristic is the single most important aspect of life on mountains, wherever they are and whatever lives on them. Zonation is the theme that accounts for the distribution of life on a mountain. Zonation is determined by temperature, which is a function of latitude, and even more, of altitude. Simply put, the higher up a mountain an organism lives, the more able it must be to cope with cold. The air temperature drops at a rate of at least three degrees F. with every thousand feet of altitude, because the thinner the air the less heat it holds.

The Influence of Latitude

Latitude, of course, modifies the effects of altitude. One must go farther up a mountain in the tropics than in the higher latitudes to experience the same degree of cold. This is evident to anyone who travels along mountain ranges that extend for great distances in a north-south direction. It can be seen clearly in the Andes, where the tree line moves progressively lower down the slopes as one goes south. Trees in the tropical Bolivian Andes ascend more than 4,000 meters up the mountainsides, while the timberline of the temperate southern Andes edges farther and farther downhill until, at subpolar Tierra del Fuego, it almost reaches sea level.

In polar regions, there is little difference between the temperatures at the bottom and top of mountains, or at least not enough to have major ecological impact. Mountains in Lapland and on Canada's Baffin Island, for example, are surrounded by tundra, which continues up to the peaks. Mountains in Antarctica are glaciated from top to bottom.

The Many Zones of Mountain Life

But mountains in most parts of the world are layered with many zones. The vegetation of the various zones is so distinctive that the stratification is visible from a considerable distance. Commonly, the zones are characterized by the overwhelming dominance of just a few types, or even one species, growing within a well-defined set of altitudinal levels. Examples of this can be found in almost any mountain system, including such a relatively low range as the Luquillo Mountains in Puerto Rico, whose highest peaks barely exceed 1,000 meters. Below 600 meters, this range is girdled with true tropical rain forest, over whose canopy tower giant tabonuco trees (*Dacryodes excelsa*), 30 meters high, with a diameter of five feet and white bark. Above 600 meters, the tabonucos disappear, the canopy is lower,

24-25. *Caribou* (Rangifer tarandus) *follow the tundra vegetation that covers many arctic mountains up to the snow line.*

and the trees are not as straight as in the rain forest. This belt gets its name from its most common large tree, the palo colorado (*Cyrilla racemiflora*), a gnarled, red-barked species that can grow to a height of 15 meters and live 1,000 years. The palo colorados are virtually restricted to an altitude of less than 200 meters; before the 800-meter level is reached they are replaced by groves of sierra palms (*Euterpe globosa*).

Some of the world's most notable plants serve as markers for mountainside zones, including species which are giants of their kind. The presence in many East African mountains of large stands of bamboo (Gramineae)—really giant grasses taller than a man—indicates that the terrain is probably above 2,500 meters. On mountains in parts of the Sonora Desert, in Arizona and Mexico, the zone between 600 and 1,200 meters can be identified by groves of towering saguaro cacti (*Cereus giganteus*), some 15 meters high, which provide a home for creatures such as woodpeckers and owls.

Another species that identifies a specific level is maquis, a low scrub which grows on the lower slopes of mountains in arid and semi-arid regions of the lower temperate zones. The scrub is dense, consisting mostly of evergreens together with small oaks. While in Europe, where it grows around the rim of the Mediterranean, the scrub is known as maquis, in the southwestern United States and Baja California, Mexico, it is known as chaparral. This term is derived from the Spanish word "chaparro," the scrub oak that mixes with the evergreens. Higher up, the scrub gives way to true forests—of pine in the Atlas Mountains of North Africa, and of belts of juniper and pinyon, ponderosa pine, and firs in the southwestern United States.

Within any given geographical area, many factors set the limits on the life zones of a mountain. Take the saguaro, for example. It needs rocky or gravelly soils. In the northern part of its range, it grows best higher up on southern slopes, where it is warmed by the sun and protected from severe winter cold. Further south, the saguaro belt is most evident on northern slopes, where it is shielded from the worst summer heat. In between, the cacti girdle the lower reaches of the mountains.

26-27. *Winds that blow westward across the vast Amazon Basin ascend the eastern slopes of the Andes and lose moisture. The heavy rainfall on the east flank of the mountains stimulates the growth of the lush cloud forests, which at upper altitudes dwindle to dwarf, or "elfin," forest. The winds lose so much moisture moving up the eastern slopes that there is little left for the cold, dry puna and altiplano, and none at all for the western side of the mountain. Virtually the only moist area on the western flank is the loma zone, where the slopes are cloaked in fog.*

The Importance of Altitude

While soil, wind, precipitation, and exposure all influence the arrangement of zones on a mountain, the basic sequence from bottom to top results from the way temperature decreases with altitude. Influenced by the increasing cold, the vegetation belts tend to narrow as they approach the top. Typically, for instance, the zone demarcated by the saguaro spans about 600 meters near the mountain base, while the belt of spruce and firs around the peak may be less than half as wide.

Not uncommonly, mountains standing in regions of barren desert can support considerable forests on their higher levels. This is because mountains, particularly on the windward slopes, intercept moisture-laden winds and steal precipitation. The higher elevations of some mountains in the Rockies of the southwestern United States, for example, receive three times as much rain annually as the deserts in which they stand.

Quite often, the slopes of mountains in the direction of the prevailing winds are lush, as are the lowlands below them, while the landscape on the other side of the mountains is bone dry. This is the case in northern Iran, where the northern slopes of the Elburz Mountains overlooking the littoral of the Caspian Sea are heavily forested and cut by surging rivers. The region on the northern side of the mountains receives about two meters of rain each year, and it falls year-round. South of the mountains, the land becomes increasingly arid, and the rains—when they fall —are seasonal; in some places less than a centimeter of precipitation occurs in a year.

At any rate, as altitude increases trees appear smaller and smaller, particularly when exposed to the wind. On the windswept ridges of many tropical mountains, this has resulted in a curious zone of forest known poetically as "elfin woodland."

The elfin woodland is a dwarf forest, where trees that are several meters tall in the forests of the lower slopes often are not much taller than a man. Twisted and gnarled, the trees of the elfin woodland are bearded with mosses and draped with air plants (epiphytes). Underfoot, the ground is springy, for it is not soil upon which one walks, but a false floor of aerial roots over the earth. The trees put out their roots above the ground in a spreading network which is covered with moss and shakes like gelatin when trod upon.

A good example of how an elfin forest develops can be found in the Luquillo Mountains of Puerto Rico, but similar examples could be cited in Africa, southern Asia, and South America. The elfin forest on the peaks of the Luquillo Mountains grows in patches among the sierra palms, which normally do best in sheltered locations. Here, they are exposed to the wind and are continually subjected to fog and mist, which wrap and twist about them. High humidity, the shroud of fog and mist, and the waterlogged soil, together with the wind, stunt the trees of the elfin woodland, preventing the formation of a rain forest like that on the slopes below.

The effects of stunting by wind, cold, and poor soil are seen most vividly in mats of dwarf trees called krummholz, or "crooked wood," which fringe the timberline on temperate mountains. On exposed slopes in the Presidential Range of the White Mountains in the northeastern United States, for example, balsam fir (*Abies balsamea*) and black spruce (*Picea mariana*), perhaps 20 meters tall in the forests below, carpet the ground in thickets less than 25 centimeters high. Where the krummholz is shielded by rocks or in low-lying areas where snow accumulates and provides insulation, the small conifers sometimes grow a bit higher, to perhaps a meter above the ground.

Eventually, if a mountain is high enough, even dwarf trees will vanish. Timberlines around the world vary considerably. In the tropical northern Andes, the tree line lies at more than 4,000 meters, while in the Presidential Range, it is at less than 1,500 meters, and in the Alps between 1,500 and 2,400 meters. The altitude reached by trees can differ even on parts of the same mountain. Where there is less wind, more snow as insulation from the cold, and more sunlight, the timberline will be higher.

Life Above the Timberline

Above the timberline lies a realm of truly alpine vegetation —meadows, moors, and, especially, tundra, scarcely

28-29. *The altitude of various mountain zones differs around the world. This chart shows some of the plants and animals typical of particular zones; like many other mountain creatures, animals often wander into other zones as well. Broken lines on the chart indicate tree lines; uneven broken lines indicate snow lines. Lighter color reflects increasing altitude.*

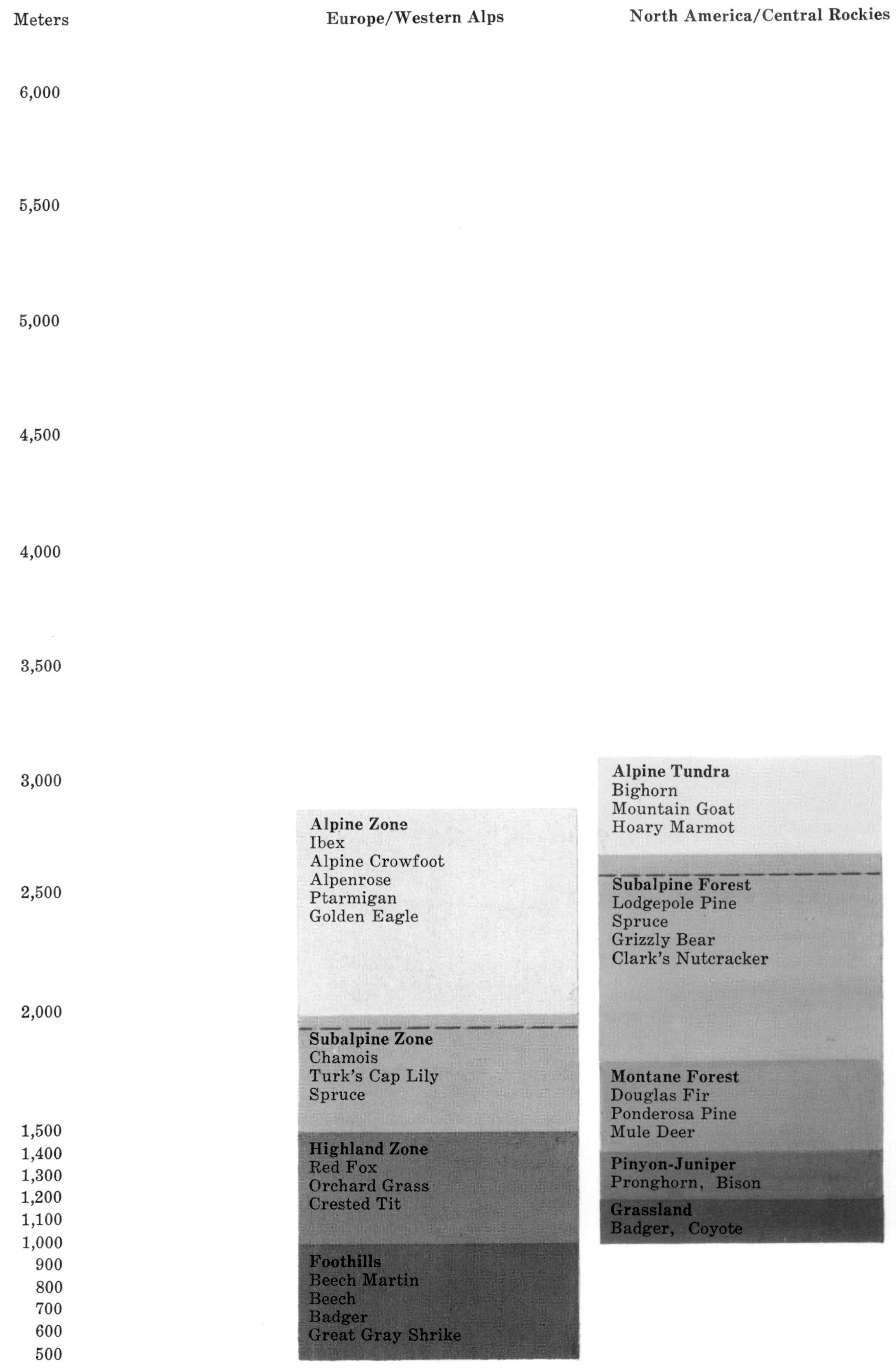

Meters	Europe/Western Alps	North America/Central Rockies

6,000

5,500

5,000

4,500

4,000

3,500

3,000

Alpine Tundra
Bighorn
Mountain Goat
Hoary Marmot

Alpine Zone
Ibex
Alpine Crowfoot
Alpenrose
Ptarmigan
Golden Eagle

2,500

Subalpine Forest
Lodgepole Pine
Spruce
Grizzly Bear
Clark's Nutcracker

2,000

Subalpine Zone
Chamois
Turk's Cap Lily
Spruce

Montane Forest
Douglas Fir
Ponderosa Pine
Mule Deer

1,500
1,400
1,300
1,200
1,100
1,000

Highland Zone
Red Fox
Orchard Grass
Crested Tit

Pinyon-Juniper
Pronghorn, Bison

Grassland
Badger, Coyote

900
800
700
600
500

Foothills
Beech Martin
Beech
Badger
Great Gray Shrike

South America/Central Andes (Eastern Side) | Asia/Eastern Himalayas | Africa/Mount Kenya

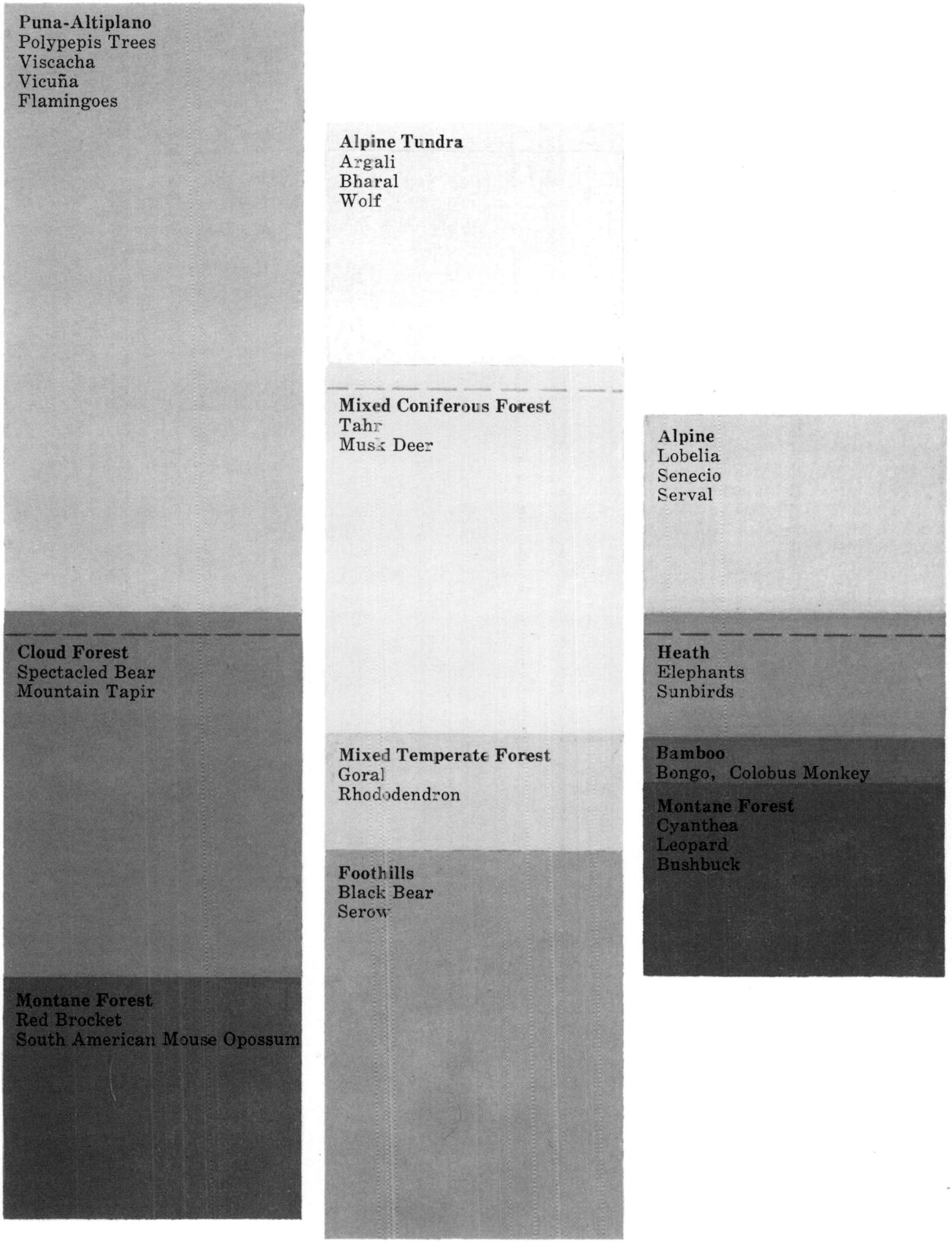

Puna-Altiplano
Polypepis Trees
Viscacha
Vicuña
Flamingoes
Cloud Forest
Spectacled Bear
Mountain Tapir
Montane Forest
Red Brocket
South American Mouse Opossum
Alpine Tundra
Argali
Bharal
Wolf
Mixed Coniferous Forest
Tahr
Musk Deer
Mixed Temperate Forest
Goral
Rhododendron
Foothills
Black Bear
Serow
Alpine
Lobelia
Senecio
Serval
Heath
Elephants
Sunbirds
Bamboo
Bongo, Colobus Monkey
Montane Forest
Cyanthea
Leopard
Bushbuck

Scales   Stoma

Transverse Section of a Leaf

Transverse Section of a Scale Enlarged

Surface View of a Scale Enlarged

distinguishable from that of the Arctic and subarctic. Rock-strewn, sprinkled with bright flowers in the spring, and dotted with bogs and clear, icy pools, the alpine zone of mountains is a harsh, demanding environment, even in the tropics. Frost touches the vegetation of the alpine belt even in equatorial Africa.

Characteristically, plants of the alpine zone must be tough, able to cope with extremes, such as fierce winds. The highest wind velocity ever recorded outside of a tornado —372 kilometers an hour—blew over the tundra of Mt. Washington, in the American Presidential Range, in April 1934.

Alpine plants have adapted in myriad ways to the short summers, long severe winters, and other harsh conditions of life in their high-altitude habitat. Many grow very slowly, only a few centimeters a year (almost all are perennials) because during the coldest months they lie dormant. To get the most nourishment and water out of the spare, rocky soil, some put out many roots. The alpine bistort (*Polygonum viviparum*), a small North American plant, reproduces by small aerial bulbs which begin to grow leaves even before they leave the parent plant, thus giving them a head start in life before the onset of cold autumn weather. The edelweiss (*Leontopodium alpinum*) of the Alps is covered with a coat of insulating hair.

Beyond the tundra, on the very high mountains, plant life diminishes to a skin of lichens and mosses on bare rock. In the central Rockies, plant life disappears at about 3,000 meters. In the Alps, plants cease growing at about the same altitude. Plants live up to about 5,400 meters in the Himalayas. If altitude is sufficient, the last zone on the mountains, above all vegetation, is one of perpetual ice and snow, a glacial cap.

Clearly, a spectacular range of environmental conditions are squeezed into a relatively short distance between the top and bottom of a mountain. Life zones that normally span thousands of kilometers in terms of latitude are compressed into several thousand meters vertically. Imagine, for example, that you are standing at the base of a mountain near the equator. About 5,000 meters above, perhaps a little higher, lies a glacial environment similar to that near the poles. To cover the same life zones horizontally would mean a journey north or south literally a quarter of the distance around the earth.

Wildlife in Migration

The vertical compression of ecozones on mountains is perhaps the single most important aspect of their ecosystems. Particularly at higher altitudes, this phenomenon creates a wide variety of habitat and food resources within a short distance for the animals that inhabit the slopes. When vegetation near the timberline of the Himalayas fails in winter, for instance, the spiral-horned markhor goats (*Capra falconeri*) descend into the valleys, where food is available in the form of evergreen oak (*Quercus ilex*).

The closely packed but sharply contrasting life zones of the mountains make it possible for alpine animals such as the chamois (*Rupicapra rupicapra*) to escape severe winter weather merely by migrating a few hundred meters downhill. The white-tailed ptarmigan (*Lagopus leucurus*) of the Rocky Mountains, for instance, moves from the

Mountain plants have developed many adaptations for living in harsh, high-altitude environments. The edelweiss (Leontopodium alpinum), above, has a fuzz of hair on its leaves. The hair insulates the plant from cold and holds in warmth. The fuzzy coat of the edelweiss also holds in moisture. Some of the rhododendrons have similar protection. One, the common alpenrose (Rhododendron ferrugineum), 30, has a covering of scales on the underside of its leaves. The scales protect pores through which moisture evaporates.

Pikas (Ochotona), above, *scurry about the lichen-covered rocks,* right, *on the heights from western North America to central Asia. Relatives of the rabbit, they are active both day and night. Pikas are known for their habit of basking on sunny rocks, even in winter. When frightened, the pika sounds an alarm call—a whistle or sharp bark.*

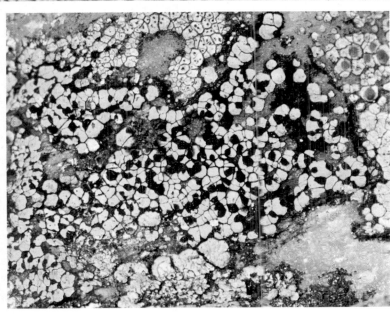

*Along the Pacific Coast Ranges,
such as the Olympic Mountains of
Washington State, heavy snows
on the slopes sometimes force
Rocky Mountain goats, top, down
to near sea level. Many mountain
animals, including the bighorn
(Ovis canadensis), bottom,
retreat downhill in winter and
return to the peaks in warm
months.*
35. *Vegetation at the timberline
is often dwarfed. Such stunted
vegetation is called krummholz.*

tundra to the timberline, where it finds food and shelter
during winter storms. With its short journey, the ptarmigan
accomplishes as great a change in ecozones as other birds
do on migrations from the arctic tundra to the north-
temperate conifer forests.

The timberline is an interface between two vastly different
communities of vegetation. At the timberline, plants and
animals of both forest and alpine belts come into contact
and interact. Because it gathers creatures from both above
and below, the timberline has an especially rich fauna and
for wildlife is one of the most critical of mountain
habitats.

Large wild animals range all the way up to the top of
many lesser mountains, and even up those of moderate size,
while smaller creatures sometimes ascend to truly dazzling
heights. Climbers more than 6,700 meters up on Mt. Everest
have discovered spiders and springtails (primitive insects)
in the glacial snows. Strong-flying birds of many species
nest far up in the heights, and, in fact, there is no mountain on earth so high that birds do not from time to time
fly over it.

Generally, however, increasing altitude diminishes the
variety of animals and cuts down on their numbers. The
ranks of creatures competing for niches on mountains are
progressively thinned as the temperature drops, but even
more so because the food supply is less concentrated up
above. One of the reasons that so many birds prosper above
the timberline is that they can cover vast areas in a short
time as they search for things to eat.

Many animals of the mountains, especially the mammals,
are the same species as inhabit the surrounding lowlands.
A variety of mammals—including wapiti (*Cervus elaphus;*
also known as American elk), brown bears, wolves, and
African elephants (*Loxodonta africana*)—regularly
ascend to the timberline, or at least very near it. Some
move into the mountains on a seasonal basis, retreating
with the approach of winter. Others wander up into the
heights to feed, then leave. Still others have found in the
mountains a last refuge, after human activities have driven
them from the lowlands. When the lowland forests of
Puerto Rico were leveled, for instance, the Puerto Rican
parrot (*Amazona vittata*) was forced into the remote
ridges and valleys of the Luquillo Range, where it barely
survives today, one of the world's rarest birds. In western
Europe the wolf still haunts a few mountain ranges—the
Pyrenees and Apennines—but hardly anywhere else.

The Miracle of Adaptation

In order to survive, high altitude wildlife must be able
to adapt. This is most conspicuous among the animals that
have managed to colonize the areas above the timberline.
They thrive on high, rugged terrain, where vegetation is
often minimal, footing uncertain, and shelter from the cold
and elements at a premium. Typical of these creatures are
the wild sheep such as the North American bighorn (*Ovis
canadensis*) and the mouflon (*Ovis ammon musimon*) of
Europe, or the wild goats, such as the ibex (*Capra ibex*).
The chief enemy confronting all such creatures—and any
others living at high altitudes—is cold. Some of them, as
mentioned, escape cold by migrating to lower elevations
in winter. Many smaller mammals, especially among the
rodents, cope with the cold by burrowing or, in the most

The tundra and crags above the timberline of the temperate zone mountains are the realm of sure-footed hoofed mammals such as the ibex (Capra ibex), *top, and the Rocky Mountain goat, bottom. Unlike the ibex, the Rocky Mountain goat is not a true goat, but a relative of the chamois* (Rupicapra rupicapra). *The ibex and mountain goat remain largely above the timberline.*

severe weather, hibernating. Other alpine animals contend with cold by means of extra-heavy coats.

The North American mountain goat (*Oreamnos americanus*), more properly described as a "goat-antelope," has a double coat, providing superb insulation, which is one of the reasons why the creature is the best-equipped of all large mammals for survival in the crags. The mountain goat can subsist on the toughest, most meager plants growing among the rocks. It is, moreover, incredibly sure-footed. Together with its Eurasian relative, the chamois, the mountain goat has a hoof rimmed by a hard, thin edge that catches on rocks and cracks, and with a sole covered by an elastic pad that serves as a tread.

South American llamas (*Lama glama*) and their wild relatives, members of the camel family, have a remarkable means of coping with another stress of high-altitude life, the decreased air pressure and resultant difficulty in obtaining sufficient oxygen to breathe. Air pressure at sea level is 6.7 kilograms per 2.54 square centimeters. But the pressure of the atmosphere rapidly decreases with altitude, so that at 6,000 meters the pressure is more than halved from what it is at sea level. This means that the higher an animal climbs, the less oxygen is pressured into its blood. The red blood cells of the llama, however, are exceedingly numerous, so that the blood has an exceptional ability to pick up and carry oxygen. The llama, therefore, gets the most out of the oxygen that is available.

The Effects of Mountain Isolation

Mountaintops are among the most isolated of natural habitats and it is hardly surprising that sometimes an animal which has managed to adapt to life on a particular mountain should remain isolated. Thus, scattered about the mountaintops of the world are a variety of creatures that are so specialized they exist at most atop one mountain mass or even on only one peak. Mt. Kenya, for instance, has its own mole shrew (*Surdisorex norae*), whose tunnels make a tracery of slight ridges in the ground of the mountain's alpine zone. Haleakala, a volcano 3,000 meters high on the Hawaiian island of Maui, has a particular type of tree geranium found nowhere else in the world. Endemism —that is, being adapted to only one locality—is more common among mountain plants than animals. Botanists estimate, for example, that 75 percent of the plants in the Guiana Highlands of Venezuela grow nowhere else.

The unique plants and animals that have evolved on mountains are the product of isolation, for, ecologically, mountains are as distinct from their surroundings as islands. Like the organisms inhabiting islands, the plants and animals of mountains can be especially vulnerable to extinction, because the habitat on which they depend is so limited. If it is destroyed, they have nowhere else to go.

Destruction in High Places

Mountain ecosystems are very fragile. The rock of mountains is often of types that are highly susceptible to weathering, such as sandstone. Given the harsh nature of mountain weather, there is a constant threat of erosion which once begun on steep slopes proceeds much more rapidly than on more level terrain and is almost impossible to reverse.

In the opinion of many ecologists, mountains are the most susceptible of all ecosystems to environmental disturbance. Unfortunately, the mountain environment is being disrupted on a monumental scale in almost every part of the globe. As never before, people have thronged into the mountains of the tropics and subtropics, which offer the last available scrap of terrain to set up homesteads, feed livestock, or gather firewood.

Some mountains—those of southern Europe, for example —felt the encroachment of man centuries ago. The impact has severely disrupted the natural balance of life. During the dry season, for instance, maquis scrub will burn like tinder, but fires are normal for this type of mediterranean ecosystem. Under natural conditions, the vegetation quickly regenerates after a fire. Greenery begins to erase the black scars of fire almost as soon as the first rains of the wet season spatter the earth. In southern Europe, however, human activity has interfered with the natural cycle of fire and regeneration. Land has been denuded and has remained that way, and the crumbly sandstone and limestone surface that has been bared has been woefully eroded. If the vegetation that remains is burned it has little chance to grow again.

In other mountain areas, which until recently were wilderness, extensive human activity is just beginning. Farms ring hillside forests where elephants roam in Africa, and there are ski resorts in some of the remote ranges of Soviet Asia. Not that all human undertakings in the heights are bad. Far from it. Parts of the Alps have had a well-managed and productive environment for centuries. For good or ill, however, modern man has affected mountain life, and has become part of the natural history of the heights.

The feet of alpine creatures such as the llama (Lama glama), bottom, and Rocky Mountain goat (Oreamnos americanus), top, are specially built for getting about on steep, slippery rocks. The llama has a flexible, elastic pad that moves and helps it get a grip. The hoof of the mountain goat has hard, sharp edges and a treaded sole for grasping hard surfaces.
38-39. The black-tailed deer, a race of the mule deer (Odocoileus hemionus), inhabits mountain forests in the northwestern United States and adjacent areas of Canada. Animals such as the black-tailed deer range up to all but the highest zones on mountains in western North America.

Europe
1. Ural Mountains
2. Scandinavian upland
3. Grampians
4. Sierra Nevada
5. Guadarama
6. Cantabrian Mountains
7. Pyrenees
8. Central Massif
9. Jura
10. Alps
11. Apennines
12. Pindus Mountains
13. Dinaric Alps
14. Bohemian Forest
15. Eifel
16. Harz
17. Ore Mountains
18. Sudeten
19. Tatra
20. Bihar Mountains
21. Carpathians
22. Transylvanian Alps
23. Balkan Mountains
24. Rhodope
25. Caucasus

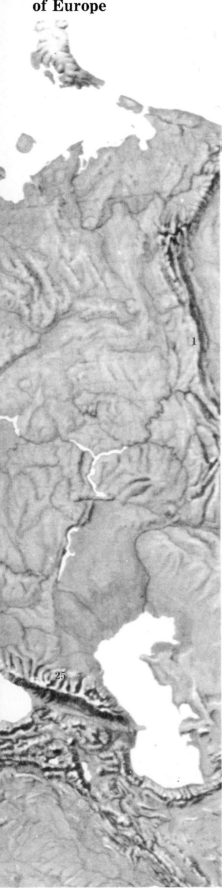

Life in the High Places of Europe

Rugged and rocky, with a crest that is permanently glaciated, the Alps rise in a great arc that begins in southeastern France, curves northeastward through Italy, Switzerland, Germany, and Austria, and then into Yugoslavia to be lost in a welter of Balkan uplands. Although by no means the only major range of mountains in Europe, and not even the highest if the Caucasus are considered, the Alps are the most celebrated European mountains. Lofty, banked with snow and iced on their summits, clad in green down below, the Alps are cut by lush valleys and capped by stony crags. Dotted with icy-clear lakes and festooned with vast upland meadows of flowers, they offer vistas stark enough to take one's breath away and yet in some places are pleasingly vibrant and evocative of peace and abundance.

The Ibexes

The Alps are truly full of life, to the rocky spires where tufts of flowers add their delicate colors to tiny ledges, and large, powerful wild goats—alpine ibexes (*Capra ibex*)— clatter up and down sheer, stony slopes. The ibex symbolizes the animals that still inhabit Europe's mountains. Many creatures living in the high places of Europe are holdouts from long ago and once roamed much more widely than they do today. Some have survived in the uplands on their own, often despite the way they were hunted down by man. Others, particularly among the hoofed animals, have been aided by human intervention and represent remarkable achievements in wildlife preservation. There is no better example of this than the ibex, which has been saved from almost certain extinction because people cared about it.

This ibex is one of several types of wild goat that range over the mountainous parts of Eurasia and northern Africa. Some varieties of ibexes inhabit the fringes of the equatorial zone, in Ethiopia. Others frequent the rocks of the Negev and Sinai, in Asia Minor, where, as in the Alps, a major effort has been made to preserve them. At home well above the timberline, the alpine ibex seldom ventures below it except in the spring when it searches for fresh sprouts to eat. During the summer, the ibex remains close to the summits, foraging among scruffy vegetation. When the winter snows arrive, the ibexes do not seek the protection of the valleys as do some other hoofed animals of the mountains, but neither do they remain near the summits. They seek a middle ground, on steep but sunny slopes, below the heavy snow of the peaks but above the timberline. The shadowed recesses of the valleys and the woods, although shielded from the wind, can be colder than higher slopes exposed to the sun. Besides, snow slides off steep slopes, while it accumulates among the trees and in the valleys. The less snow cover, the better for the ibex, which must be able to get at the low vegetation on which it feeds.

As a general rule, even when there is no snow cover, ibexes shun the valleys. Their populations thus tend to be scattered and separate on the crests of mountain ramparts, although they can be distributed across the high places of an entire mountain system. Such is the case in the Caucasus, where the two local varieties of ibex—which, unlike the alpine type, venture into the forests—remain plentiful. Before the Middle Ages, the alpine ibex, like those

of the Caucasus, was widespread, even if scattered. But even then the process that was to bring it to the brink of extinction had begun.

There is something about animals like the ibex that arouses admiration and even envy in the human breast. Perhaps it is because the ibex seems so wildly free and unfettered, leaping among crags where few if any people can go. To people in medieval times, the ibex represented powers far beyond those of humans. Ibex horns could be fashioned into charms, and other parts of the creature were believed potent against all sorts of diseases. The blood, for example, supposedly removed calluses. Balls of hair, stones, and other materials that remained undigested in the stomach, and the heart muscle, were considered general cure-alls. Not surprisingly, although laws were enacted to protect it, the ibex was hunted incessantly.

Before the end of the Middle Ages, the alpine ibex was in trouble. By the sixteenth century it was rare in Switzerland and in the Tyrol. The last record of the goat in the Tyrol was in 1706, when a herd of five bucks and seven does was reported. By early in the last century, the only remaining alpine ibexes inhabited the southern slopes of the Italian Alps, in the Piedmont. At this point, however, the fortunes of the alpine ibex changed, largely because the royal house of Sardinia and the Piedmont valued the goat as game. Increasingly well-policed laws were created to protect the last ibexes, and in 1854, King Victor Emmanuel II established a reserve, Gran Paradiso, which made the territory inhabited by the ibexes royal property.

Royal game wardens were stationed in the mountains to protect the ibexes, which gradually began to multiply in their haven atop the Piedmontese Alps. The population reached 600 animals by 1879, and before the beginning of World War I had passed the 3,000 mark. Meanwhile, the ibex had been reintroduced into other parts of the Alps from which it had vanished, and therein lies a tale of international intrigue.

Hoping to repeat the success of the Italians with the ibex, the Swiss in 1892 founded a reserve, the Peter and Paul Game Park, planning to re-establish a population of the goats with specimens bought from the Italians. The Italians, however, would not sell any of the Gran Paradiso ibexes, and the only other ibexes available were some hybrids from the crossing of zoo specimens with domestic goats. The Swiss opted for direct action: in 1906 they captured a few of the ibexes in the Gran Paradiso, and smuggled them over the border into Switzerland. This ploy brought about Italian cooperation, and additional ibexes were sold to the Swiss.

Since then, the ibex has been successfully reintroduced in several portions of the Alps. France, Switzerland, and Austria as well as Italy now have thriving ibex populations. Despite the losses of some ibexes in the two world wars, about 8,000 of them now live in the wild, and, moreover, quantities of them have been bred in zoos.

The alpine spectacle has been immeasurably enhanced by the return of the ibex to the heights. In most of its range, the ibex is the largest surviving wild animal, some males weighing more than 120 kilograms and standing 75 centimeters high at the shoulder. Its brow is crowned by massive horns, comprising a tenth of the body weight and curving slightly backward to sharp points.

44-45. *The ibex (Capra ibex) is a wild large goat that lives above the timberline in many European mountains. When engaged in combat for dominance and the right to mate with the females, male ibexes cross horns and thrust against one another, or sometimes rear up on their hind legs and clash horns. Each will try to engage its opponent from higher ground so that it may have the advantage in a pushing match.*

46. *The green lizard* (Lacerta viridis) *is one of several related types found all across Europe, including the mountain regions. Males engage in savage combat, with legs stiff, bodies arched, and heads low. During the ice ages reptiles were pushed south of mountain barriers such as the Alps and the Pyrenees. In Mediterranean lands, the group to which the green lizard belongs underwent extensive diversification. Most of the green lizards are egg layers, but one species* (Lacerta vivipara) *bears living young.*

During the rutting season in winter, rival males battle
one another, hooking and hammering their magnificent
horns against one another but seldom causing injury. The
horns of the female are less imposing, but can be wielded
with deadly effectiveness in defense of the young. Although few predators large enough to threaten the alpine
ibex remain within its range, the golden eagle (*Aquila
chrysaetos*) occasionally will swoop from the sky and
seize a kid.

The ibexes of the Caucasus, on the other hand, still must
contend with several dangerous mammalian predators.
The brown bear, wolf, lynx, and leopard all will dine on
ibex if the chance arises. The Caucasian ibexes generally
resemble the alpine species, but have horns with a more
emphatic curve. So does the ibex of Spain (*Capra
pyrenaica*), found in scattered populations on mountain
ranges throughout the Iberian peninsula.

Like the ibex, most of Europe's other larger and more
imposing animals inhabit the mountains, but a few big,
splendid creatures, such as the wisent (*Bison bonasus*),
still exist in isolated lowland sanctuaries such as the
Bialowieza Forest, which straddles the border between
Poland and the Soviet Union. It is truly remarkable that
so many spectacular animals have survived in Europe's
mountain fastnesses, adding interest and excitement to a
continental fauna limited by nature and impoverished
by man.

The natural restrictions on the fauna, and also the flora,
of Europe are due mainly to the small size of the continent,
which covers only about 6,200,000 square kilometers.
Strictly speaking, of course, Europe is not a continent at
all, but rather a huge peninsula, jutting westward from
Asia. Exactly where Asia ends and Europe begins always
has been a question. Generally, however, geographers now
concede that Europe is largely walled off from Asia by two
mountain ranges: the Urals directly to the east, and the
Caucasus, between the Black Sea and Caspian Sea, in the
southeast.

Except for its arctic fringes, all of Europe's land area
lies in the North Temperate Zone. Europe, in fact, is the
only continent except Antarctica without a portion in the
tropics. Its plants and animals are therefore limited to
those which have adapted to the upper latitudes.

During the Pleistocene ice ages, when glaciers reached
into central Europe, the north-to-south range available to
the life of the continent was further compressed. Warm-
climate trees such as the tulip and sweet gum were gradually pushed out of Europe. Many forms of life were caught
in a squeeze between ice advancing from the north and
the great mountain barrier—formed by the Pyrenees, the
Alps, and the Carpathians—that divides central Europe
from its southern tier. The advance of the ice sheets halted
a considerable distance north of the mountain wall, but
between the ice sheets and the mountains, boreal forest,
cold steppe, and tundra replaced the temperate, deciduous
forest. Creatures of the temperate forest were forced
south of the mountain barrier—that is, if they could
surmount it. If not, they perished. Meanwhile, from the
mountain heights, local ice caps bore down over the adjacent
lowlands, further compartmentalizing the living space
remaining for temperate plants and animals. When the
ice receded, as it did periodically in the Pleistocene era

*Shown here is the farthest
extent of the last two Pleistocene
glaciations of Europe. The
hatched area marks the Riss
glaciation, and the blue area
indicates the Würm, last of the
ice sheets. During glacial times,
when ice sheets reached from
the north into the European
heartland, the glaciers on mountains such as the Alps, Pyrenees,
and Caucasus, well south of the
main ice area, also moved down
the slopes.*

Races of the fire salamander (Salamandra salamandra) are spread through much of Europe. This spotted fire salamander (Salamandra salamandra terrestris) inhabits the Jura Mountains of Switzerland. The name of the fire salamander is derived from an old folk belief that the creature was born in fire. The belief stems from the fact that the salamanders were seen creeping from under the bark of logs being burned in fireplaces.

(and has been doing for about the last 12,000 years), some of the refugee species managed to get back over the mountains and recolonize their old range.

The Adaptable Salamanders

The results of the interaction between the glaciers and mountains can be seen quite clearly in the distribution of certain salamanders. The European fire salamander (*Salamandra salamandra*), for example, needs the shade and moisture of temperate forests such as beech woods. This species in Europe has split into eight subspecies, easily distinguishable by contrasting patterns of yellow markings on their black bodies. Six of the subspecies inhabit southern Europe, each restricted to its own peculiar area, while two reach well up into the center of the continent. Scientists believe that the fire salamander was driven south by the ice sheets, then spread north again when the ice receded. During its exile, the salamander occupied isolated refuges along the Mediterranean, where Europe's last deciduous forests remained. While in isolation, the species differentiated. When the climate warmed, the various groups of fire salamanders began to spread out once more. Some interbred, creating yet other varieties. Today, the diversity of the fire salamanders is manifest not only in a great range of markings, but also in different body sizes within the same species. The Corsican fire salamander (*S. salamandra corsica*), for example, is speckled, while the Italian fire salamander (*S. salamandra gigliolii*) is covered with blotches running most of the length of its body. Similarly, the Pyrenean fire salamander (*S. salamandra fastuosa*) is slight and slender, while the more southerly Spanish fire salamander (*S. salamandra bejarae*) is robust.

When the last ice sheets retreated, two subspecies of fire salamander crossed back over the mountain barrier into their old ranges. One, a banded variety, came from the southwest, either from Spain or northern Italy. The other, a spotted type, moved in from the southeast, where it is found around the curve of the Mediterranean to Israel. Where the salamanders found moist, deciduous forest, as in the foothills of the Harz Mountains, they proliferated.

Another type of movement in response to glacial advance was carried on by the alpine salamander (*Salamandra atra*), a glossy black creature that haunts the heights from the western Alps to Albania. Living even above 3,000 meters in altitude, this salamander was probably driven down from the mountain slopes when the alpine glaciers expanded, and found haven along the Mediterranean and Adriatic seas. Later, when the climate moderated, the salamander ascended the mountains again. In Spain, the small gold-striped salamander (*Chioglossa lusitanica*) seems to have been thrust out of the Pyrenees into the northwestern section of the Iberian peninsula, but for some reason was unable to return to its old home when the glacial period ended. To this day, the salamander still inhabits its ice-age haven.

With the moderation of climate and the retreat of the glaciers to the north and back up into the heights, many animals which had inhabited Europe during glacial times either followed the ice sheets or vanished altogether. The true elk (American moose—*Alces alces*) moved north with the boreal forest belt of conifers, birch (*Betula*), alders

(*Alnus*), and aspens (*Populus*), and is no longer found in
the south-central part of the continent. As post-glacial
forests spread over the open country north of the mountain
barrier, the horse (*Equus caballus*), adapted to steppes,
retreated to the north and east.

Thousands of years later, the decrease in the range of the
horse would have great implications on the history of
human civilization. It was not in western Europe but on
the plains of southern Russia, or thereabouts, that the horse
was first domesticated. This creature, which during the ice
ages had been one of the main sources of food for prehistoric Europeans, returned to the western part of the
continent, pulling the chariots of marauding Indo-
European warriors from the east. It was largely through
the military supremacy made possible by the horse that the
Indo-European peoples were able to displace, or absorb,
the earlier inhabitants of Europe and lay claim to the
continent.

In the Wake of the Glaciers

During glacial times, the steppe and tundra of western
Europe, north of the Alps and the Pyrenees, were also the
home of the alpine marmot (*Marmota marmota*), chamois
(*Rupicapra rupicapra*), and ibex, today typical of the
high mountains. As the glaciers crunched down the slopes,
they pushed the high-altitude animals before them, ever
downward, until mountain creatures found themselves in
the lowlands. By then, however, the lowland climate had
become as severe as that of the peaks had been, so the
mountain animals found the steppe, tundra, and boreal
forest a suitable home. When the ice returned to the
heights, the chamois, ibex, and marmot followed in its
wake to the alpine zones they occupy today. Plants also
colonized new highland habitats that opened up when the
ice inched back to the summits. The process of coloniza-
tion continues today on some peaks of the Alps where
glaciers recede and new plants from below are gaining
footholds.

Many times, as glacial periods alternated with interglacials,
the plants and animals of the European mountains north
of the Mediterranean rim have moved up and down the
slopes. The rim of the Mediterranean has been more stable.
Only the very highest peaks have been glaciated, and thus,
for the most part, the plants and animals of the mountain
ranges in southernmost Europe are the product of unin-
terrupted development in an environment which remained
relatively undisturbed until the pace of human activity in
the region began to pick up about 2,000 years ago.

What Man Has Wrought

Although it has occurred within only a few ticks of the
geological clock, the change wrought by man on the environment and life of European mountains has made them vastly
different from what they were only a few centuries ago.
Human activity has not always resulted in catastrophe,
of course, and in some places it has even been in harmony
with nature; but in most cases, man has brought a devasta-
tion to the mountains and to their plants and animals.
Nowhere has the face of the mountains been disfigured
more than around the Mediterranean, where a rich, fruitful
wilderness has been turned into a landscape that looks
blasted and lifeless. In ranges such as the Apennines of

Alpine meadows and tundra blaze with multicolored wild-flowers in the spring. Many flowers have recolonized the heights as the glaciers have re-treated up the peaks. Many of the high meadows and heaths of the Alps are natural. Over the centuries, however, logging and grazing by domestic flocks and herds have created new meadows and in some places lowered the timberline. Row 1: left, *bear-berry* (Arctostaphylos nova-ursi) ; center, *shrubby milkwort* (Poly-gala chamaebuxus) ; right, *Koch's stemless gentian* (Gentiana kochiana). Row 2: left, *yellow alpine pasqueflower* (Pulsatilla alpina apiifolia) ; center, *spiny thistle* (Cirsium spinosis-sinum) ; right, *orange lily* (Lilium bulbiferum). Bottom, *oxlip* (Primula elatior).

Italy, the Sierra Nevada of Spain, and the fabled hills of Attica, in Greece, the same sad story of waste and destruction has been repeated. The forests have been stripped from the slopes with no thought to the regeneration of the trees. Year after year, domestic sheep and goats have been herded higher and higher into the hills to devour every scrap of vegetation and trample into the dust what is not eaten. Without the trees, there has been no source of humus to enrich and build the topsoil. Without the cover of vegetation, the topsoil itself has been stolen by wind and water. Lacking the stabilization of trees, the slopes have deteriorated, and water sluicing downhill unimpeded by grass or tree roots has left great scars in the hillsides. This sort of erosion on mountain slopes feeds on itself, growing worse each time it rains, particularly on crumbly sedimentary rocks such as those of the Mediterranean area. The result of centuries of such erosion is in many places a landscape that looks as barren as the surface of the moon.

Although nature has given way in the face of the onslaught, scraps of wilderness still can be found here and there among the coves and on the peaks of the Mediterranean mountains, giving at least an impression of what the landscape once was. Not far from Rome, in the central Apennines, for instance, some of the higher slopes remain covered with forests of stately beech trees. Some wolves live in this region, as well as brown bears, which have vanished from most of western Europe. The pardel lynx (*Lynx pardellus*), tassel-eared and long-legged, still hunts in the Sierra Morena Range of southern Spain, where islets of vegetation have managed to return to land that was denuded long ago.

The maquis that is typical of the lower altitudes of Mediterranean mountain areas has in some places crept upwards over slopes from which the old, deciduous forests have been removed. Although far different from the green, shadowed woodlands of earlier times, the scrubby maquis nevertheless is a wild habitat in its own right. Tangled, nearly impenetrable in many places, the maquis shelters a variety of animals. On Corsica and Sardinia, where bandit gangs have traditionally made their strongholds in the maquis, it is also the haven of the diminutive Mediterranean race of red deer (*Cervus elaphus*), which is nearly extinct, and the feeding grounds of the mouflon sheep (*Ovis ammon musimon*) from the higher rocks.

Man in the Alps

The hand of man has also transformed the Alps, but in a gentler manner, at least up until the last few decades. The second-highest European range—the Caucasus is higher —the Alps are rugged, craggy and sheer, but they have been infiltrated by a network of broad, verdant valleys, excavated by the glaciers which moved down from the peaks during the Pleistocene era. Through these valleys, many of which cradle sky-blue lakes or clear surging rivers, people have penetrated into the heart of the Alps, and farms, villages, and cities are spread throughout all of the range, which covers more than 124,000 square kilometers.

Far below stony heights where ibex cavort and the fragile-looking but tough edelweiss clings to tiny ledges are bustling communities and prosperous homesteads, a world seemingly infinitely removed from the peaks. From the

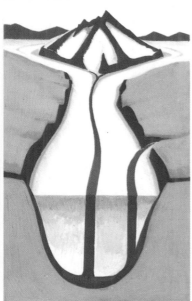

valleys, however, human influence has spread up the mountainsides, chiefly in the form of farming and forestry. As in other European mountains, the Alps above the tree line have remained relatively untouched. There has been no wood-cutting because there are no trees, and grazing pressure has been minimal. Below the timberline, however, the Alps suffered some of the same pressures as the mountains farther south, except that political and economic circumstances averted an environmental disaster of the magnitude of that in the Mediterranean.

The greatest damage to the environment of the Alps occurred before 1850. Most of the population of the region was poor and agrarian, while at the same time much of the land was owned by a limited number of wealthy people. For subsistence, on the one hand, and riches, on the other, the mountains were subjected to excessive exploitation. Slopes were cleared of trees, sometimes by burning. Cattle and other livestock were permitted to overgraze the pasturage, timber was carted away to be used for building material, fuel, and charcoal.

The disappearance of autocratic economic systems, and the development of a trading economy, with a subsequent increase in the standard of living, halted the massive exploitation of the Alps a century ago. At the same time, legislation protecting forests, streams, and pastures was introduced, so that by the middle of this century man and nature had reached a compromise. Scrupulous management of the environment created a balance between farm, pastureland, and woodlot, and the natural mountain ecosystems. Much of the vegetation of the Alps, admittedly, is not natural, but nevertheless it is green and pleasant. Moreover, especially at higher altitudes, some natural forests remain, along with alpine meadows enlivened by bog whortleberries (*Vaccinium*) and primroses (*Primula*), and glades in which grow wild gardens of rhododendrons (*Rhododendron*)—all of which harmonize with the meadows and tree plantations created by man.

Within the past few decades, however, the boom in tourism in the Alps, as in other high temperate mountain systems, threatens the environmental balance that has been so carefully maintained. Urbanization is spreading out from large mountain resorts, disrupting not only the physical environment, but also the human living patterns that have contributed to the equilibrium. Lured by profits in the tourist industry, for instance, some farmers have left their meadows unmown. The tall grass becomes heavily encrusted with snow in the winter. When the snow cover melts in the spring, it creeps slightly downhill, dragging the heavily-laden grass up by the roots. The unmown grass, furthermore, provides a slick surface that smooths the way for avalanches. The danger of these damaging snow slides is also increased by ski slopes, which are, in effect, man-made paths for avalanches.

Pyrenees and Caucasus

The Alps, so heavily imprinted with the stamp of man, contrast dramatically with the Pyrenees and Caucasus, which support true wildernesses. Here, the evidences of human activity are few and far between—a lonely track here, a tiny village there, or perhaps a flock of sheep tended by a herdsman and his dogs. To understand the lack of human presence in these two mountain ranges, one must look back again to the Pleistocene era. Because of their more southerly location, the Pyrenees and Caucasus did not spawn such massive glaciers as the Alps. The valleys of the southern ranges, therefore, were not so extensively excavated by the ice as those of the Alps and so to this day remain narrow and sheer-sided, not easy to traverse and not particularly inviting to people.

The absence of human impact upon the Pyrenees and Caucasus is reflected in their vegetation. It looks wild and unkempt, not manicured like the forests and fields of the Alps. Parts of the Caucasus, which is more than 1,200 kilometers long and 180 kilometers wide, are clad in heavy forests the likes of which are only a memory in most of Europe. In the valleys and on the lower slopes, especially in the moist western reaches of the range, grow vast woodlands of mixed deciduous trees—oaks, elms, and beech, for example. Further up, beech predominates, then birch and maple, and finally, up to the timberline, spruce. These forests are the realm of the bear and the wolf, the wildcat and even the leopard, the wisent and the chamois.

The Chamois

The chamois (*Rupicapra rupicapra*) share the high peaks of the Caucasus, the Alps, and many other European mountains with the ibex. Lacking some of the bulk and weight of the ibex, the chamois is perhaps even more graceful than the big goat. Skipping about the crags, leaping across chasms more than five meters wide, or perched with all four feet upon a tiny ledge, the chamois seems to appear and disappear like a phantom. One minute a patch of grass beneath a talus slope can be empty, and the next minute the tawny form of a chamois can be seen resting on the ground. Was the animal there all along? Or did it slip by unnoticed?

Much more common than the ibex, the chamois resides not only in the Alps, but in ranges from Spain to the Caucasus, and north to the Tatras of Poland as well. Chamois move about the Pyrenees, the central Apennines, the mountains of the Balkans, and the Carpathians, often in substantial numbers. The young are born in the spring, up among the grasses of the highest alpine meadows. Usually the chamois has only a single kid, but in rare instances as many as three young may be born. Within a few hours of birth, the young chamois is able to amble about with its mother, staying very close to her heels. If for some reason the mother dies, the kid may be adopted by another female chamois.

During the summer, the young and females gather in herds which can number more than a score of animals. They sometimes can be seen trooping across the skyline, traveling single file, negotiating dizzying spires and walls while foraging for juicy herbs and alpine flowers. In summer the adult males rove alone, staying away from the herds of females and young. Autumn brings the rutting season, and the males seek out the herds and compete with one another for the right to mate with the females.

Unlike the rutting battles of many horned, hoofed creatures, those of the chamois are not ritualized to avoid serious damage to the participants. Unlike the spectacular but harmless clash of various mountain sheep, chamois males engage in a deadly fencing match. The weapons are slender horns, up to about 200 milimeters long, with a pronounced hook near the tips. The combat may rage up and down the

54. *Glaciers have reshaped mountain valleys.* Top. *Here a river valley is shown prior to glaciation, while it is still shallow and gently sloped.* Center. *A glacier descending from the nearby peaks scoops out the valley, scours rock and soil from the earth, and builds up moraines along its edges.* Bottom. *After the glacier retreats, the valley is deep, sheer, and littered with debris. The riverbed descends steeply with many waterfalls.* Below, The Matterhorn, 4,478 *meters high, towers above Zermatt, in Switzerland, which nestles in a valley that forms a green cleft in the Alps. Valleys gouged out by glaciers during the Pleistocene era have enabled man to penetrate into the heart of the Alps.*

56-57. *A graceful denizen of the heights, the chamois (*Rupicapra rupicapra*) sometimes engages in bouts of leaping and running about, perhaps simply to rid itself of excess energy. Chamois live on the uppermost slopes as long as they are bare of snow, descending only during severe winter weather. The chamois ranges from Spain to Asia Minor. This agile creature has a knack for seeming to appear and disappear like a ghost.*

slopes, for, since the clash is not ritualized, it does not end when one of the contestants turns and flees. Rather, when one of them breaks off the fight and attempts to get away, the other may pursue ruthlessly.

When severe winter weather ravages the higher elevations, especially if the snowfall is heavy, the chamois descends below the timberline into the forests, where it finds food in the form of young pine growth, lichens, and mosses. The upper levels of the forests have lost some of their hoofed summer tenants by the time the chamois arrives, for changing seasons bring a vertical displacement of the creatures inhabiting the various zones of the mountains. With the coming of winter, many animals migrate down into the next zone. Just as the chamois descend into the conifers, the deer move from the evergreens into the deciduous woods below. In summer, however, the deer ascend far up into the conifer zone, even to the timberline. In the Swiss Alps, the red deer (Cervus elaphus) commonly move up through spruce groves to an altitude of almost 3,000 meters. In the Alps, Carpathians, Caucasus, and similar ranges throughout the continent, the roe deer (*Capreolus capreolus*), the most common deer in Europe, adds grace and elegance to the higher slopes.

The Adaptable Deer

These two species of deer typify the way in which many large mammals—particularly among the hoofed herbivores and some of the predators which follow them—have adjusted to mountains, although, unlike the chamois and ibex, they are not specially adapted for life in the heights. Both deer are animals of the forest. The roe deer, particularly, thrives where woodland is broken by meadows and fields. Where the deciduous forest has been cleared from the lowlands, but still persists on the slopes, the roe deer has gone with it. If the lowlands around the forested mountainsides are occupied by farms, the situation is then ideal for the red deer, which has the best of both worlds— a place of shelter in the forest and a superb food supply in the farmers' fields.

The adaptability of the deer tribe has been emphatically demonstrated by the red deer of the Grampians and other parts of the Scottish Highlands. Well over 100,000 of the big deer roam the rugged, fog-swept moors that now cover the mountains. The deer herds, some containing hundreds of animals, range over a treeless landscape covered by heather and moorgrass and studded with boulders and rocky outcrops. The moorlands carpet even the crests of the mountains, a thousand meters high.

Formerly, the moors were not nearly as extensive as they are today, and certainly not the primary feature of the highland landscape. From their original location on exposed heights, the moors have expanded downward because of human activities, such as clearing of forests and subsequent overgrazing by sheep. The invasion of the lower slopes by the moors has been largely checked, in recent years, but the forest has not returned. Yet the red deer, which has relied on the forest since prehistoric times, has managed not only to survive but to prosper.

The Return of the Wisent

The wisent, or European bison (*Bison bonasus*), the most awesome of all European hoofed mammals, is like the red

Neck swollen, antlers polished, and looking for a battle, a European red deer stag proclaims his readiness to mate.
60-61. The red deer of Europe is one of several races of the species Cervus elaphus, which ranges across Eurasia and North America, and even is found on the fringes of northwestern Africa. Included in this group is the maral of Asia and the wapiti of America. Like many other large hoofed mammals, the red deer is often a creature of the mountains as well as the lowlands.

Above. *The mountains of Corsica, rugged and wild, along with the mountains of Sardinia, are the native haven of the mouflon (Ovis ammon musimon). The mouflon is native to Corsica and Sardinia but has been transplanted successfully to mountainous regions throughout much of Europe. It is a hardy, prolific animal that is able to exist on a wide variety of plants, even the poisonous deadly nightshade (Atropa belladonna).*

63. Like other mountains of southern Europe, those of Corsica have been almost totally stripped of their original forests. In many places, however, this thick scrub called maquis has overgrown the lower slopes. At times such as when snow caps Corsica's Sierra Nevada Range, the mountains still seem remote and wild.

deer partly an animal of the mountains, specifically, of the wild Caucasus. But it would not have survived in the Caucasus without direct human intervention. The Caucasian wisent, the southernmost of the two races of wisent, seems to have been quite widespread in earlier times. The Greek soldier of fortune Xenophon told how the Paphlagonians of Asia Minor used drinking horns, which were probably taken from the wisent. Much later accounts, from early in the seventeenth century, describe the animal as ranging through most of the Caucasus, especially the northern and western portions. The days of free-roaming wisent herds were numbered, however, both in the Caucasus and the rest of Europe. Destruction of forests and ruthless hunting pushed the lowland race into the backwaters of Poland and Lithuania, and the Caucasian animals higher and higher up the ridges. World War I swept over all of the remaining bison refuges, and by the 1920s, none of the animals were left in the wild. The last members of the Caucasian herd are believed to have been shot by Russian revolutionary soldiers, apparently for political reasons, for the wisent had nominally been the property of the Russian czar.

A few wisent from both races, however, remained in zoos, and today some of their descendants have been reintroduced into reserves. The wisent in the Caucasus, where the Soviet government protects them, number in the hundreds. However, they are not counted among the total number of surviving wisent because when the Soviets built the herd after World War II, the shortage of breeding animals prompted them to cross the Caucasian animals with American bison (*Bison bison*). Even so, the return of the wisent to their ancestral mountains should be welcomed. There are few sights in the animal kingdom more thrilling than seeing wisent traveling through the deep forest. Their huge but rangy bodies move with astonishing stealth and lightness. A dozen of the shaggy beasts may pass nearby, yet all that may be heard is an occasional rustle of leaves, or rarely, the snap of a small twig. As they feed in the shadows of the boughs, weaving between the trunks of the trees, the wisent evoke images of what Europe must have been like in that very distant time when people were few and large wild animals populated the length and breadth of the continent.

During the late Pleistocene era, the mountains of southern and central Europe were populated with a small wild sheep, whose brown hide was marked with a distinct white saddle, especially when in winter coat. This sheep, the European mouflon (*Ovis ammon musimon*), has survived in historic times only on the islands of Corsica and Sardinia but has been hunted so intensely there that today the island populations number only in the hundreds. Luckily, the mouflon is an extremely adaptable animal and has prospered in the many other European mountains to which man has transplanted it during the last century and a half. There are several thousand mouflon living in Germany, and thousands more in Austria, Czechoslovakia, Hungary, Rumania, Poland, and other nations. The European mouflon has become so widespread that, according to some estimates, the total population of this sheep has reached 20,000 individuals.

One reason for the mouflon's success as a transplant is that it can feed on a great variety of plants—from soft grasses and herbs to the young growth on trees. Nimble and fleet, endowed with enviable climbing ability, the mouflon is very much a mountain creature, but not of the highest crags; instead, it haunts the upper reaches of the forests and, especially on its native islands, it can also be found in the maquis. Typically, mouflon do not pick their way among stony areas, but canter lightly over the rocky ground. They are extremely wary and exceptionally keen of vision and smell, but once frightened, they can be driven into a trap. Traditionally, Corsican and Sardinian hunters have driven the sheep into passes where ambushes are set for them. Although hunters have decimated the mouflon population in the islands, they have been largely responsible for the survival of the species, for it was mainly their interest in new species of game that led to the introduction of the mouflon into its new homes.

War on the Predators

Typically, when wild animal populations have been decimated, as they were in the mountains of Europe, the first animals to benefit from human intervention are hoofed mammals such as the mouflon and ibex. The reasons for this are easy to understand. In the first place, the hoofed mammals are not direct competitors of man, although they do compete with the livestock of farmers and ranchers for forage. Where herds of hoofed mammals are immense, with their members counted in the millions (as was the case on the plains of eastern Africa and western North America), people have eliminated them to make room for stock. But the mountains, especially at higher altitudes, support nowhere near the number of herbivores as the plains, so they generally have not offered competition to livestock. Moreover, truly alpine herbivores such as the ibex and chamois live where domestic animals, even sheep and goats, seldom venture. Secondly, hoofed mammals pose no direct threat to humans, although when provoked, some of the horned or antlered creatures—mountain goats, for instance —can be fierce adversaries. At the same time, such animals are valued as game, and thus have been considered worth saving.

Predators, however, because of their position in the natural scheme of things, are in a much less favorable position with regard to man. Some of the larger predators occasionally prey on humans, and most of them, from big cats to stoats (*Mustela putorius*), will not hesitate to kill and eat man's domestic animals if they have the opportunity. Furthermore, merely in fulfilling their natural roles, predators compete with man by killing the same game animals hunted by people. Thus, particularly in rural or agricultural areas, the interests of predators and people can conflict most emphatically. Not surprisingly therefore, people, even while actively engaged in wildlife conservation, have waged endless war on the predators. It was not too many years ago, for example, that 150 wolves were shot in the Bialowieza Forest of Poland with the approval of the same forestry and conservation administrators who were overseeing the restocking of the wisent there. Only within the past few years, in fact, have Polish biologists begun to call for protection of the wolf, which had been regarded as vermin, to be destroyed by any means.

This state of affairs is not peculiar to Poland among the nations of Europe. Recognition that large mammalian predators merit the same type of preservation efforts as the hoofed mammals is just beginning in many countries. In view of this fact, the mountains have taken on even more importance as havens for and reservoirs of wildlife, for in much of Europe except for the far north, it is only in the high places that the bigger predators such as the wolf remain.

Sleek, Gray Shadow

The image of the wolf (*Canis lupus*) that has roamed the landscapes of the human imagination is much more often than not a ravening horror, a baneful enemy of man, a sleek, gray beast always lusting after blood. Superstition, folklore, legend, and ignorance—and rarely actual rampages by wolves—have contributed to the wolf's low reputation, which is almost totally undeserved. It is because of the fear and revulsion that wolves inspire, rather than their occasional depredations upon livestock, or their even rarer attacks on people, that these gray hunters have been exterminated from most of their range.

The wolf once inhabited almost all of Eurasia, from India to the Arctic snowfields. Although still common in many parts of central and northern Asia, the wolf has disappeared from most of Europe. The creature is extinct in eleven European countries—Ireland, Great Britain, France, Belgium, the Netherlands, Denmark, Switzerland, Austria, Hungary, and both of the German states. Conservationists consider the wolf as good as gone from Scandinavia, although there are hopes that the species can be reintroduced there in the future by release of wolves bred in zoos.

Small wolf populations survive in the Iberian peninsula, Italy, Bulgaria, Czechoslovakia, and Poland, as well as in the European part of the U.S.S.R. The wolf remains rather common in the tundra of northwestern Russia, the Urals, the mountains of the Balkans, and especially the Caucasus, which has a wolf population just about as dense as that of any other part of the globe. More than 4,000 wolves roam the Caucasus Mountains and their environs, a region which in some places has a density of one wolf per 20 square kilometers of territory. During the winter snows, wolves in search of food even venture into the suburbs of the few large towns of the area.

It is not hard to understand why the people of the Caucasus might be skeptical about pleas for protection of the wolf as an endangered species. Although the red deer, roe deer, wild boar (*Sus scrofa*), and similar natural prey abound in the Caucasus, wolf packs occasionally descend upon domestic herds and flocks. The Soviet government, in fact, considers the wolf population of the Caucasus too large for the welfare of the area's cattle breeding efforts, and carries on a systematic program of wolf control there, to the chagrin of some conservationists in other countries.

Wolves of the Carpathians, and in the mountains of Balkan countries such as Yugoslavia, were subjected to extensive programs of organized eradication after World War II. The animals, which had been mainly restricted to the mountains for more than fifty years, multiplied and began to move out of the highland wilderness and down into villages and towns during the turmoil of the war, fulfilling an old Balkan belief that the wolf is 'war's brother.''

Wolves, found in Europe and North America, can adapt to many types of habitat, from tropical desert to arctic tundra. Although they are not basically mountain animals, in many places mountains provide a refuge for them.

By the middle of the 1970s, the intensity of the extermination programs had been moderated, and the goal of outright elimination of the wolf was replaced by an attempt to manage the species in balance with game and livestock.

Hunters of the Carpathians

The wolf of east-central Europe and the Balkans is primarily a creature of the deciduous forests, and therefore of the mountains. If large tracts of such forest still covered the lowlands, the wolves might survive as well down below as they do, for instance, in the heights of the Carpathians.

An offshoot of the Alps, the Carpathians arc for almost 1,300 kilometers through eastern Europe, with the Tatra Mountains at their northern extreme, and the Transylvanian Alps of Rumania edging their southern tier. The bulk of the Carpathian Range is in Rumania, which mainly because of its ownership of the mountains still can claim 40,000 square kilometers of wolf habitat, roamed by perhaps 2,000 of the swift-hunting beasts.

Much of the southern Carpathians are covered with vast, undisturbed forests. The woodlands of the lower slopes are pure beech (Fagaceae), while higher up the beech trees mix with Norway spruce (*Picea abies*), followed in turn by larch (*Larix europaea*) and stone pines (*Pinus cembra*). Approaching the 2,000-meter mark, the conifers begin to appear noticeably shorter, and on the higher mountains, merge with alpine meadows, enlivened in the spring by the blooms of azaleas (*Loiseleuria procumbens*) and rhododendrons (*Rhododendron kotschyi*).

Wolves in the Carpathians, and in the Balkans as well, tend to migrate up and down the mountainsides according to the season. During the summer, when herdsmen drive cattle, sheep, and goats into the higher mountains, the wolves follow. This is the season when the depredations by wolves upon livestock are most severe. When the snow piles up on the peaks—in some parts of the Carpathians the accumulation is six meters deep—the domestic herds are brought down from the mountains. The snow also drives the wolves to lower levels, where they hunt their natural prey, deer and wild boar.

Haven in the Apennines

A southern leg of the Alps, the Apennine mountain chain fills the Italian boot to its toe. Although of modest height—the highest peak, Gran Sasso d'Italia in Abruzzi, is less than 3,000 meters—the Apennines are quite rugged. Despite their location in the heart of a region peopled for centuries, these mountains still contain pockets of wilderness sufficient to support small populations of wolves, the last in Italy.

Possibly ten distinct wolf populations are scattered down the Apennines, ranging southward from the knee of the boot. Numbering perhaps a hundred animals in all, the various groups seldom if ever meet, because they are separated by deep valleys heavily populated by man.

As in the Carpathians, the wolves in the central Apennines are in close contact with domestic livestock, especially in the summer and fall, when sheep are driven up the slopes. Some of the sheep fall prey to the wolves, and occasionally

the damage is significant. One night in the autumn of 1976, for example, a small wolf pack swept into a pen holding 400 sheep and killed 67 before shepherds awakened and came to the rescue. In the southern Apennines, cattle replace sheep as the main type of livestock and there, since wolf packs number no more than two or three animals, losses are considerably lower than farther north.

Extensive studies have been made of the wolves living in the area of the Abruzzi National Park not far from Rome, as part of a joint effort by the Italian government and the World Wildlife Fund to rescue the Italian wolf from the threat of extinction. Biologists discovered that in winter, the wolves—less than two dozen survive there—hide in the thick beech forests that in some places reach more than 1,800 meters up the mountainsides They generally stay in the forests and keep away from the human inhabitants, who for the most part live either in towns and cities in the valleys, or hamlets perched on some of the lower mountains. Once darkness cloaks the mountains, however, the wolves, in packs of up to seven animals, slip out of the forests to prowl around the villages. Sometimes the big beasts—the European wolf weighs about 50 kilograms—even venture to the fringes of large cities, including Rome. Generally, however, they restrict their nocturnal forays to the mountain villages, where they raid garbage heaps, or snatch up chickens, pigs, dogs, calves, and colts if these are found in the open.

The visitations of the wolves to the villages and towns of the Apennines demonstrate that wolves can exist amidst territory heavily populated by people if a forest haven is nearby. Moreover, the wolves must be given well-policed protection, as has been done in Italy. In 1976, the Italian government decreed that all wolves were to be permanently protected. At the same time, to avoid raids by wolf packs on the mountain villagers' livestock and the consequent retaliation against the wolves, an effort is being made to restock the Apennine forests with red deer and roe deer, providing natural prey for the wolves and adding new variety to the natural history of the mountains.

Brown Bears

Remarkably, the same Abruzzi forests also support several dozen brown bears (*Ursus arctos*), perhaps as many as sixty of the huge, shaggy creatures It is one of the few concentrations of brown bears in Europe outside the U.S.S.R., although until the Middle Ages the bear inhabited the length and breadth of the continent. Today, the range of the species in western Europe coincides exactly with the location of mountain ranges that still contain relative wilderness. In addition to the Apennines around Abruzzi, the bears still roam in the Carpathians, the Pindus of Greece and adjacent ranges in Albania and Yugoslavia, the Scandinavian Range, the Pyrenees and the Cantabrian Mountains, just west of the Pyrenees.

In all of the Alps, however, the only permanent population of bears lives in the forests below the towering limestone spires of the Dolomites in Italy's Trentino—Alto Adige region. Only a very few bears remain there, and lacking the protection of a national park such as those in Abruzzi, they are almost certainly doomed. Increasingly, the

Wolves in central and southern Italy occupy well over 8,000 square kilometers, but because these regions are separated by valleys and human settlements, different species of wolves seldom mix. Once widespread through Europe, the gray wolf (Canis lupus) *now survives west of the Soviet Union only in a few mountain areas. The only considerable wolf populations on the continent outside the U.S.S.R. are in the mountains of the Balkans and in the Carpathians. Pockets of wolves survive in the Italian Apennines and in some mountains of Spain and Portugal.*

forests are being cleared to make room for ski resorts and other developments, and while the Dolomite region retains its magnificent mountain scenery, it is losing its wilderness.

Occasionally, a few bears from Yugoslavia stray into the easternmost part of the Austrian Alps, but they are either driven out or killed. A few years ago, for example, four or five bears were sighted in the vicinity of the Rax Plateau, southwest of Vienna. Newspapers in the area demanded the bears be destroyed for fear they would endanger children on their way to school.

Although it belongs to the same species as the savage grizzly (*Ursus arctos horribilis*) of North America, the brown bear of Europe is a rather docile creature. Rarely, and only under unusual circumstances, does it attack any other large animal or humans. Despite its weight, which may be more than 250 kilograms, the brown bear seldom kills animals larger than marmots, although it will readily eat the carrion of sizable creatures—deer, for instance. Even in the Caucasus, where bears thrive, they seldom attack domestic animals, despite the fact that a full-grown brown bear can kill even a steer with ease.

Generally the bears depend on plants for food, particularly in spring, summer, and early autumn. Plenty of plant food is available to the bears in the deciduous forest that is their prime habitat, and in the conifers in some of the more northerly areas which they roam. Lush mountain valleys, where the soil is enriched by minerals and nutrients carried down by erosion from the slopes, offer particularly fine foraging, but during the warmer months the bears do not hesitate to climb beyond the tree line to search the meadows for the likes of alpine lettuce (*Lactuca alpina*) and bilberries (*Vaccinium myrtillus*).

For the most part, adult bears rove alone, except in late spring and early summer when courting and mating occurs. Once a pair of bears has mated, each of them goes its own way. For the male, the reproductive process has ended for the season. The female, on the other hand, has just begun the job of continuing the species. She carries her young for almost as long a gestation period as a human mother. With the coming of the cold weather and snow, the bear heads for a den, which may be in a cave, an excavation dug into the earth, or a crevice amidst the roots of a great tree. In these snug winter quarters, the bear goes to sleep, but not, as commonly believed, into the comatose state that characterizes true hibernation. Fat with the berries, grubs, and small rodents garnered by nearly incessant feeding during the autumn, the bear slumbers, but its body temperature does not drop as it does in creatures that really hibernate. The bear can be roused by the approach of an intruder, or even a spell of exceptionally warm weather can bring it forth from its den. The young, one to three in number, are born while their mother overwinters. Within the shelter of the den she nurses them, until spring is under way and the family can leave and begin the search for food.

Cats, Large and Small

Bears and wolves are not the only large carnivores ranging the mountains of Europe. Within the vastness of the Caucasus the leopard (*Panthera pardus*) prowls, not

in great numbers, but nevertheless often enough to qualify as part of Europe's fauna. The presence of the leopard on the fringes of Europe should not be surprising. No other cat has adapted to a wider geographical and ecological range, although the mountain lion of the New World comes close. Leopards range almost the entire continent of Africa, and, where they have not been exterminated, throughout Asia except for the frigid northernmost regions. Mountains, near-desert, jungles, forests, swamps—almost any sort of habitat suits this most adaptable of all big cats. Although we tend to associate the leopard with the tropics, it stalks its prey as far north as the southern fringes of Siberia.

After the leopard, which may weigh more than 90 kilograms, the largest wild cat of Europe is the lynx (*Felis lynx*), which weighs about 10 kilograms. Long-legged, with tassel-topped ears, and gorgeous fur of sandy to reddish-gray color, the lynx, like the wolf and the bear, is a forest animal which has been driven into the wildest of mountains. Outside the Soviet Union and Finland the European lynx survives only in mountains—the Carpathians (especially the Tatra Range and the Transylvanian Alps), the Scandinavian Range, the most isolated of Balkan peaks, and the Caucasus.

The lynx preys mainly on ground birds, hares (*Lepus capensis*), and small rodents, but it can also bring down animals as large as deer, especially when such prey are mired in the snow, over which the lynx travels on its broad, padded feet. In Poland, lynx have reportedly killed even elk (called moose in America), although probably only the young, aged, or infirm.

Similar to the European lynx, although slightly smaller, is the pardel, or Iberian, lynx (*Lynx pardellus*), which once inhabited most of the Iberian peninsula and parts of southern France. Today the pardel lynx is still found in the recesses of the Pyrenees, the Sierra Morena, and the mountains of Toledo, as well as some lowland areas in the estuary of the Guadalquivir River.

Hardly larger than a tabby, but famed for fearlessness when cornered by man or dogs, the European wildcat (*Felis sylvestris*) has in most places retreated into the mountains like its larger relatives. The only wildcats left in Britain, for example, remain in the Scottish Highlands, especially on the windswept moors of the Grampian Mountains. Wildcats are still fairly common in most of the European mountains that shelter the larger predators, and because of small size, even survive in a few lowland forests.

Extremely solitary by nature, the wildcat is much less conspicuous than a larger feline such as the lynx. Moreover, the wildcat needs considerably less area for a hunting range, and smaller prey. Although it can kill grouse, hares, and even larger animals such as roe deer fawns with ease, the wildcat can get along exceedingly well on squirrels, mice, voles, and even fish. Thus today, the wildcat still pads through the Ardennes, and the lowland scrub of remote areas in Spain, Corsica, Sicily, eastern Europe, and the Balkans.

The distribution of the wildcat today, mostly in the mountains but still partly in the few forested regions remaining at lower altitudes, reflects an earlier stage in the history of the larger mammalian predators. While the wildcat is

Mountains are among the last
habitats of two of Europe's
largest wild cats, the lynx
(Felis lynx), top, and the pardel
lynx (Lynx pardellus), bottom.
Like other large predators, these
tassel-eared felines have become
extremely rare. Outside the
Soviet Union and Finland the
European lynx is extant only in
mountainous regions. The pardel
lynx, the less common of the two,
exists almost only in Spain.

70. The mountain hare (Lepus
timidus), top, and the capercaillie
(Tetrao urogallus), bottom, are
typical of the creatures that are
the prey of cats like the lynx.
If the number of hares declines,
a decrease in the lynx population
follows a year or so later.

Once ranging as far north as
Norway and Sweden, the feisty
wildcat (Felis sylvestris) has
been pushed south to the southern
tier of Europe, except for a few
northern havens such as the
Scottish Highlands. In the latter
area, especially in the Grampian
Mountains, the wildcat has
prospered. It is a deadly hunter
of rodents, birds, rabbits and
similar small creatures. The wildcat is known for its ferocity,
especially in defense of its young.

considered rare, its range is much wider than that of any other predatory mammal in Europe except for the ubiquitous red fox (*Vulpes vulpes*) and some of the weasels and their kin (Mustelidae). In view of the fact that at long last attitudes toward predators are beginning to change, and their role in keeping nature balanced is being recognized, perhaps the wildcat will be able to repopulate some of the areas from which it has vanished.

Predators of the Air

The predator that is most characteristic of the peaks of Europe is not a mammal but a bird—the golden eagle (*Aquila chrysaetos*). So majestic that in medieval times only kings were allowed to use it for falconry, the golden eagle favors the crags as a nesting site, although it may also inhabit sheer sea cliffs, and in the Asiatic part of its range will venture over open country as well.

Like other birds of prey, the golden eagle has suffered at the hands of man. It has been shot by hunters and farmers who fear it will take their young livestock. Its eggs have been stolen by falconers. It has been poisoned by pesticides which have hindered its reproductive processes, and the wilderness solitudes it demands have been increasingly disturbed by lumbering, tourist development, and other human activities.

Yet the golden eagle survives. Soaring on wings with a span of more than two meters, this master of the air can be seen over the Grampians, Apennines, Carpathians, Pyrenees, the Pindus, and similar ranges. About 100 pairs breed in the northern part of Sweden, and a few pairs still nest in the rugged brown hills of Sicily. In the Alps it is the only remaining predator of creatures such as young chamois and ibex, although it subsists mainly on hares and marmots. Above the tree line of the Alps, on the open meadows, one can find scattered here and there the remains of marmots eaten by eagles: A few bits of tattered fur, perhaps a small bone or two, testify that an eagle has struck from above.

When hunting, the golden eagle flies low over the mountainsides, taking a sector of slope at a time, eyes searching the landscape for marmots, hares, or game birds such as grouse or ptarmigan. For such a large bird—more than five kilograms in weight—the golden eagle is remarkably maneuverable in the air. It can follow the twists and turns of even a fleeing hare with impressive agility. Once on target, the eagle rockets down upon its prey, talons spread, and slams into the victim at a speed of more than 140 kilometers an hour. The impact upon the prey is almost always fatal.

The nests of the golden eagle are perched on tiny ledges and the stony faces of the most inaccessible crags. It is often difficult to perceive how the huge constructions of twigs and sticks built by a breeding pair of eagles can be affixed to what appear to be virtually sheer cliff faces. When the cupped interior of the nest holds eggs or young, the parents line it daily with fresh greenery. Usually two eggs are laid, but sometimes the number is twice that.

After an incubation period of about forty days, the eggs hatch. For three months—spanning the summer—the eaglets are bound to their lofty nest. They are a demanding brood, continually squealing for food, which is brought to them by both parents.

74 top. *The lammergeier, or bearded vulture* (Gypaetus barbatus), *is a creature of the most remote crags. It is known for its habit of carrying animal bones aloft and then dropping them to break on the rocks below. The bird then eats the marrow of the shattered bones.*
74 bottom. *Monarch of the mountains of the Northern Hemisphere, the golden eagle* (Aquila chrysaetos) *is the only large predator left in many montane regions, such as the major portion of the Alps. Rodents and ground birds form a large part of the eagle's diet, but this large bird will also try to take a young chamois.*

77. *A large bird that inhabits both mountains and plains, usually far from human habitation, the black vulture* (Aegypius monachus) *can be found in Spain, some of the Mediterranean islands, and the Balkans. It nests in trees and on cliffs.*

78-79. *The griffon vulture* (Gyps fulvus) *has a distinctive silhouette in flight, different from that of other vultures in its range. Its wings are very long and broad. Its tail is extremely short, squared and dark, and its head is extremely small and almost hidden by its neck ruff.*

The Great Vultures

Almost as magnificent as the golden eagle are the larger vultures which inhabit the mountains of the Iberian peninsula, Greece, the Balkans, and a few locations in between. Although vultures generally eat only carrion, one, the bearded vulture, or lammergeier (*Gypaetus barbatus*), with a range that touches the Mediterranean lands of Europe, is said to occasionally prey on small mammals and even knock chamois off the edges of cliffs and ledges. It is a white-headed bird whose golden underparts make a colorful sight as it sweeps overhead.

Strictly a mountain-dweller, the bearded vulture is essentially an Asian and African species, but it also lives in the Pyrenees, Corsica and Sardinia, the Balkans, the Caucasus, and occasionally in Sicily and the easternmost Alps. Although its predatory habits make the bearded vulture unusually interesting, it has another mode of feeding that is even more curious, and that has earned it the name of "bonebreaker." It feeds mainly on the marrow of bones, generally from the carcasses of carrion. Once the other types of vultures have dined, the bearded vulture alights near the carcass, and begins to sort out the scraps, searching for bones. Smaller bones are swallowed by the bird, while larger ones are carried aloft to immense heights above the rocks. Then the vulture drops its burden, so that the bones shatter on the rocks, exposing the marrow.

Two other exceptionally large vultures nest in some of the mountains along Europe's southern fringes. They are the black, or hooded, vulture (*Aegypius monachus*) and the griffon vulture (*Gyps fulvus*). The black vulture builds its nest in trees or on cliffside ledges in the Iberian peninsula south of the Pyrenees, in Sicily and Sardinia, and across the southern Balkans to the Caucasus. The griffon has a similar range, but edges a bit northward toward the European heartland. It sails high over the Pyrenees, the Maritime Alps, and parts of the northern Balkans, as well as in most areas occupied by the black vulture.

From a distance, the two vultures appear to resemble one another. They are among the largest of vultures, with bodies more than a meter long. Unlike most other vultures, they both have very broad wings, rather like those of the largest eagles. Both are brownish, but the black vulture is darker; up close, the tawny highlights of its plumage are clearly visible. Around its neck, each of the vultures has a ruff of feathers that looks like a feather boa.

In behavior, the birds contrast. The griffon is a sociable sort of bird, roosting with its fellows, and nesting on ledges and caves with others of its kind. But the black vulture is very much of a loner, shunning its fellows except during the breeding season. Both feed on carrion, and the black species often haunts village garbage dumps in search of refuse from human kitchens. The griffon approaches its food with great caution, staying away from carcasses until all signs of life have long ceased. Despite its imposing size, the griffon has a reputation as a coward, and it will waddle quickly away from its meal even at the approach of creatures as small as a village cur.

The Raven and the Eagle

To some people, vultures seem ghoulish, however graceful they appear as they soar and glide through the air. They

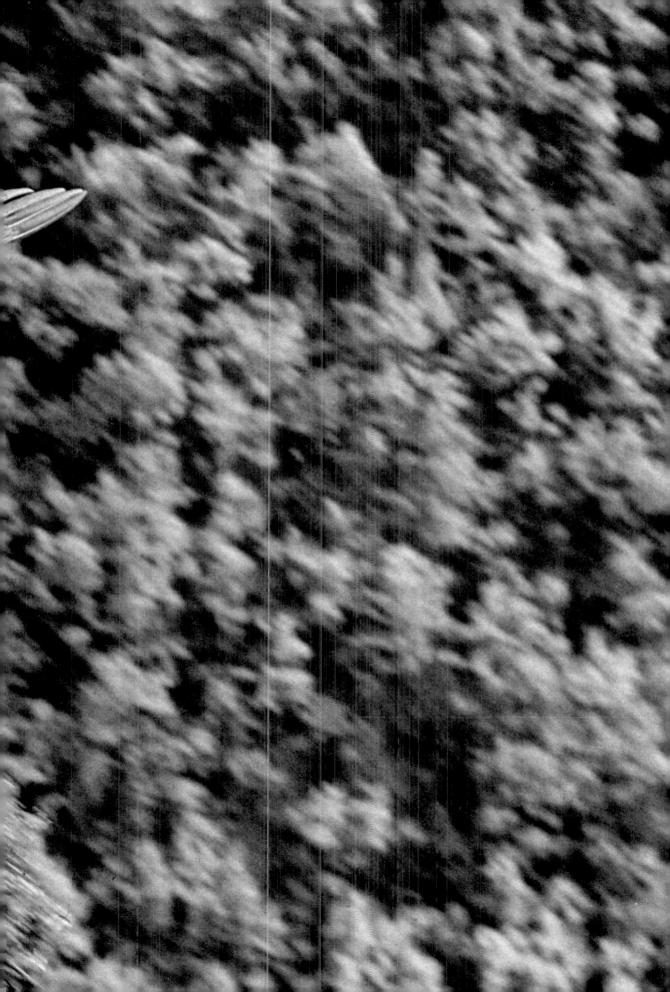

Above and right. *The imposing griffon vultures, seen here with the carcass of a domestic ram in the Pyrenees, usually gather in large groups to nest on ledges and in caves on cliffs. They often set up housekeeping only a few yards apart.*

81. *Where there are mountains or cliffs, ravens are almost always there, nesting among rocks, or soaring high above the peaks. Tough and aggressive, ravens (Corvus corax) are distinguished from crows by their large size and shaggy feathers on the throat. They make large nests of sticks, twigs, grass, moss, and other plants. A female raven generally produces a clutch of between three and eight eggs, which she incubates without help from the male. The eggs, greenish-gray in color, hatch in about 3 weeks. Young ravens are ready to fly in little more than a month. Ravens have an extensive range throughout Eurasia, North America, and south to Mexico, Arabia, and northern Africa.*

are, of course, important members of nature's sanitation force, which cleans up the leavings of other creatures, including man. The vultures are assisted in this task by the black roughneck of the air, the raven (*Corvus corax*). From the tropics to the Arctic, the raucous croak of the raven can be heard over a great variety of landscapes, but nowhere does the raven seem more at home than among the bleakest rocks and crags. Ravens inhabit all of Europe's mountainous areas, ceaselessly patrolling in search of carrion. But they are also predators in their own right and kill a wide variety of small mammals. Given the opportunity, ravens will eat almost anything, from insects and fish to acorns.

Bulky of body, ravens are nevertheless aerial acrobats, especially in the breeding season. They circle, loop, and soar, tumbling and rolling over in the air. As part of what seems to be a courtship display—which creates such strong bonds that ravens mate for life—the male raven can even fly upside down, a remarkable feat of aeronautics.

Another avian predator of the mountains that is capable of dazzling flight is the Bonelli's eagle (*Hieraaetus fasciatus*), which inhabits the middle and lower slopes of ranges in Mediterranean countries, except on the Italian mainland. With narrow wings and a body about 66 centimeters long, this eagle resembles a large falcon. Streaking through the air, its shrill cry echoing over the cliffs on which it nests, it dives down on its prey from great height, in a heart-stopping swoop almost as spectacular as that of the peregrine falcon (*Falco peregrinus*). This eagle, which rarely strays north of its usual range, is a predator of small mammals and birds.

The Small Mammals of the Mountains

The European mountains abound in the small mammals which constitute the prey for feathered and furred hunters. Although they are seldom evident, mice, voles, shrews, and similar small, four-legged creatures scurry and skitter about the floor of the mountain forests and over the alpine meadows and moors. Very often, these little animals spend much of their time under the ground, in burrows. Beneath the surface of alpine and subalpine meadows on the main Caucasus Range, for example, lives the small but stocky burrowing vole, or Prometheus mouse (*Prometheomys schaposchnikowi*). Peculiar to these meadows, the vole excavates a complex system of burrows, radiating outward from a grass-lined nest at its center. As the tunnels spread out from the nest, they come nearer to the surface but do not open above the meadow grasses, for the inhabitants of the burrow systems prefer to remain underground. The only times that the burrowing vole comes to the surface, in fact, is when the tunnels dug with its long, curved claws have been disrupted, or when it cannot find enough food down below, the roots of herbs being its main diet.

Another interesting vole inhabits mountains in Yugoslavia at altitudes between 680 and 2,100 meters. It is the snow vole (*Dolomys bogdanovi*), first identified as a fossil species in 1898, and then discovered as a living animal a quarter century later. This grayish rodent emerges at night from cracks and crannies in the limestone of peaks such as Mt. Trebevic to forage for grasses and other vegetation. It stores food in its nests for the winter. The fossil from

which the snow vole was first identified was discovered in Hungary. Since the discovery, other fossil snow voles have been found in many parts of Europe, demonstrating that the snow vole is another creature which had a wide range in prehistoric times, but now inhabits only a small mountain region.

March of the Lemmings

Prone to sudden and rather mysterious increases in numbers, the Norway lemming (*Lemmus lemmus*) is the most famous of all the small creatures in the mountains of Europe. This snub-nosed rodent, the size of a small rat, occasionally concentrates in immense throngs and then sets out on a heedless, frantic march to nowhere in particular—a journey which may end in lakes, rivers, the sea, or merely dispersal over the countryside.

The lemmings occupy the upper reaches of Scandinavia's mountainous spine. Stout little creatures, whose ears are so tiny they are hardly visible, they generally remain on lichen-covered heathlands well above the timberline, even above the scrubby thickets of willows that mark the limit of tree-like vegetation. The lemmings bustle about the year round, spending much of the summer beneath the thick mats of lichens and mosses, which help shield them from predators such as weasels, owls, stoats, and ravens.

When the snows of winter cover the landscape, the lemmings do not migrate to lower altitudes like many of their enemies; they shift their nest sites to drier ground, away from excessive dampness that would threaten the survival of their young.

Hidden in their tunnels under the snow, lemmings may produce several litters of from three to nine young each. While the parents forage beneath the snow—an activity that coincides with an enlargement of the flat claw on their "thumbs"—the young nestle in nests of moss, lichens, and other soft plant materials. If the winter is long or the weather fickle, many of the young and their parents will die. However, a short winter, with stable weather but also enough snow to provide cover from predators, promotes lemming survival. Once the snow melts, swarms of lemmings may emerge, and zoologists then record another population explosion of the small rodents.

Sometimes, lemmings multiply so rapidly that their population overflows from the alpine zone. Occasionally, the mood of such dispersals becomes frenzied, and the lemmings undertake their famous marches. The reasons for the marches still are debatable. It may be that once lemmings reach a certain concentration within a given area, the stresses of crowding triggers headlong flight. Because in many parts of Norway the mountainsides slope almost directly into the sea, some of the marching lemmings do rush into the waves and die. Others perish in fresh waters or are snatched up by the gathering of predators that harries the rodent travelers. But many of the lemmings also survive the marches, settle down, and resume their normal activities.

The Long Sleep

Unlike the lemmings, the hefty alpine marmot (*Marmota marmota*) of the Alps and Tatras avoids the snows that sweep the high ground above the timberline. The marmot makes its den in a long burrow, which can reach to a

Voles (Microtus) *of many varieties inhabit mountain meadows and rock piles. They undergo sharply defined population cycles.*

Small but fierce predators, the pine marten (Martes martes), above, and the weasel (Mustela nivalis), left, are marked by lightning speed, ferocious determination, and needle-sharp teeth. The weasel can squirm through openings the size of a coin to root out small rodents in their dens. The marten, a forest dweller, is a marvelous climber and the terror of arboreal squirrels.

84-85. The alpine marmot (Marmota marmota), a large ground squirrel, is a favorite food of large birds of prey, especially eagles. Piles of marmot hair and skin left in alpine meadows testify to the success of the hunting birds. If found in the open by an enemy, the marmot streaks for its burrow.

depth of about three meters beneath the surface. When the spring and summer sun warms the alpine meadows, however, the marmot is very much in evidence on the surface. It is a diurnal creature, feeding on alpine plants and sunning itself luxuriously when it has the chance. The sight of a golden eagle overhead will send it scurrying for its burrow, of course, but unless danger threatens, the marmot—really a large ground squirrel—stays out of its den for most of the day.

The approach of winter changes the marmot's way of life. It retires to its den, well below the frost line, and hibernates. Its slumber is of the deepest sort and in fact represents true hibernation, with body temperature and metabolism operating at a level just high enough to keep the animal alive. The marmot remains dormant for up to eight months, until the arrival of the spring rouses it.

Color Changes with the Season

For some creatures of the high mountain zones, the coming of winter is heralded by a change in coloration of their fur or feathers to white. Among these animals is the mountain, or blue, hare (*Lepus timidus*), which is common in the Scottish Highlands and ranges south from Scandinavia into the Alps. As winter approaches, this hare molts its brown coat and becomes bluish-white, matching the rocks and snow.

The change in the plumage of the rock ptarmigan (*Lagopus mutus*), a bird related to the grouse, is an even more dramatic example of natural camouflage altering with the seasons. The change is gradual and goes on almost all year long as the seasons fade into one another. During the winter the ptarmigan is snow white, except for slight red and black markings near the eye, and a black tail, which is concealed when the bird is resting. By summer, only the wings and underside of the bird are white. The back of the bird is mottled black on brown; by autumn it is gray.

The ptarmigan is a rugged bird, hardy enough to remain above the tree line except during the most severe blizzards. In Europe (the rock ptarmigan also inhabits northern Asia and North America), it can be seen in much of Norway, both the upland and lowland areas. Elsewhere it is strictly a bird of the mountains. It roves the high meadows of the Alps and Pyrenees, and the moors of the Scottish Highlands, where it has shown its adaptability by thriving even amidst ski resorts.

Birds from Top to Bottom

Just below the Highlands realm of the ptarmigan lives its close relative, the red grouse (*Lagopus lagopus*). In summer, the two birds are difficult to tell apart, except that the plumage of the red grouse has a slightly ruddy hue. In Norway, the red grouse is known as the willow grouse because it roams the willow thickets edging the lower boundary of the ptarmigan's zone. The willow grouse also lives in the vast birch forests in the valleys of the Scandinavian "Alps," and in North America where it is called willow ptarmigan.

A similar type of zonation of game birds occurs in the Caucasus, where the snow partridge (*Tetraogallus caucasicus*) inhabits the alpine belt, and the Georgian

Ptarmigan (Lagopus) *are known for the way their plumage changes with the seasons. All ptarmigan are white in winter, as is demonstrated by the rock ptarmigan* (L. mutus), *above. The willow ptarmigan* (L. lagopus), *86 top, and L. mutus, 86 bottom, are shown as plumage is changing.*

black grouse (*Lyrurus mlokosiewczi*) stays in the rhododendron thickets and the birch and conifer forests just
below the timberline. The boundary between the habitats
of the two birds, which are found only in the Caucasus, is
sharply defined by altitude. The partridge seldom strays
below 3,000 meters, except during the worst winter weather,
and the grouse hardly ever goes above that level.

Because of the different zones, most European mountains
have a profuse variety of birds. Within a few hundred
meters, one can find not only mountain birds but also those
of the lowlands. The lowland varieties range well up into
the foothills of many mountain ranges. In the beech forests
at the base of the Alps, for example, the great gray shrike
(*Lanius excubitor*) hunts mice, smaller birds, and insects;
and the middle-spotted woodpecker (*Dendrocopos medius*),
with its crimson crown and white shoulder patches, makes
its nest holes in the highest parts of the beech trees. At
lower levels of the Pyrenees, the bright, rosy underparts
of the bullfinch (*Pyrrhula pyrrhula*) can be seen among
the greenery, and sometimes the liquid song of the willow
warbler (*Phylloscopus trochilus*) can be heard. Higher up
in the Pyrenees and the Alps, deep within crevices on cliffs
between 1,800 meters and the snow line, the wall creeper
(*Tichodroma muraria*), marked by its sickle-like bill and
crimson wings, nests and rears its young. Among the same
bare precipices lives the snow finch (*Montifringilla nivalis*),
which vacates the barren rock above 1,800 meters only when
the storms of winter swirl among the stony spires. Above
them all sweep the great birds of prey, and the smaller
cousin of the raven, the alpine chough (*Pyrrhocorax
graculus*), which ridges the updrafts and high winds
endlessly, as though for the joy of it.

Far Travelers
During the autumn migration, the avian spectacle of the
European mountains is most splendid, particularly around
the passes through which the migrants penetrate the highland barriers between northern Europe and the milder
lands to the south. Fleeing the chill of the approaching
winter, the flying travelers move south in an endless stream,
funneling through the gaps in the mountain wall. Along the
passes of the Austrian Alps, one can watch flock after
flock of finches, distinguished by the dipping pattern of
their flight as they sweep low over the alpine meadows.
Chaffinches (*Fringilla coelebs*), in flocks of one or two
dozen, stream through as though bouncing on the air,
coursing from the heights into the green valleys. The
crossbills (*Loxia curvirostra*) travel in somewhat larger
flocks, usually about a score of birds. Occasionally, beckoned
by the conifers they seem to love dearly, the crossbills stop
to rest. The brick-red males lend a touch of color to the
dark green boughs of the spruces close to the tree line, as
though they had been decorated for Christmas.

Barriers
Passes can cut through the physical obstructions of the
mountains, but ecological barriers are not so easily surmounted, especially for animals which walk, crawl, or
fly close to the ground. The higher the pass, the fewer
the animals that can use it, because the environment becomes colder and less productive with altitude. Many of
the animals of Sweden, for instance, have been barred

Unlike the common chough
(Pyrrhocorax pyrrhocorax), the
alpine chough (P. graculus),
above, seldom, if ever, leaves the
high mountains. The common
chough sometimes nests on sea-
side cliffs as well as in the
mountains.

88 top. The mountain-dwelling
rock thrush (Monticola saxatilis)
ranges through southern and
eastern Europe. A shy bird, it
flits quickly into the shelter of
rocks and boulders if startled.
It often sings from perches on
rocks.

88 bottom. Truly a bird of the
heights, the alpine accentor
(Prunella collaris) is found in
the mountains of central, southern, and eastern Europe, but not
in surrounding lowlands. It nests
in holes and crevices among
boulders and low brush.

90-91. Bramblings (Fringilla
montifringilla), common in the
birch forests on Scandinavian
mountains, migrate southward as
far as central Italy and Greece
in the winter. The brambling is
one of many finches that travel
through the alpine passes on their
way to wintering grounds.

from neighboring Norway because of the Scandinavian mountains, which separate the two nations. However, several species have penetrated the barrier across low-lying passes. Almost two dozen different types of ground beetles (Carabidae), for instance, have bridged the Scandinavian mountains through the lower, forested passes. Higher up, the migration of the ground beetles becomes more difficult, as is demonstrated by the fact that only eleven species have made it across the belt of spruces and firs, and only seven have crossed the subalpine zone.

The expansion of the range of the ground beetles in northern Europe has been going on for about 9,500 years, ever since the glaciers that covered most of Scandinavia retreated. Meter by meter, the little beetles, many of which cannot fly, have taken over ground vacated by the ice. Since the last ice age, a small, creeping insect called the bag-worm moth (*Solenobia triquetrella*), the females of which are flightless, has made a similar invasion of the Alps. The moth, which lives under rocks and in the chinks of tree bark, can barely crawl. Yet it has managed to colonize all of the Alps to the very edges of the glaciers, and is still moving into new ground as the glaciers continue to shrink. Since 1874, for example, the Rhone Glacier, at the river's headwaters, has retreated more than 1.5 kilometers. The bag-worm moth has followed right behind the ice. The insect was collected in the 1960s on ground which had been covered by the glacier only 40 years before.

In the millennia since the end of the last ice age, many reptiles have repopulated central and northern Europe by moving back over the barriers of the Alps and Pyrenees from refuges in the south. The viperine snake (*Natrix maura*), for example, has crossed the Pyrenees from Spain and extended its range through southern France, and into the Alps as well. This water snake, which feeds on fish, follows stream beds up into the mountains, and has been observed at altitudes above 2,000 meters.

Varieties of Change

The movement of animal populations across the mountains, and up toward the heights, demonstrates the dynamism of life in the high places of Europe. Today, as in the past, the numbers of some species are swelling, while others are decreasing. Some animals and plants are conquering new territory, while others are losing ground. When due to natural events, such fluctuations are normal. Nature is always changing, sometimes quickly, sometimes with the slowness of ages, often imperceptibly, but occasionally with catastrophic impact. Organisms must adapt to such natural changes or perish. That is the way of nature. It is only when man interferes with nature and alters the environment by his activities that change is unnatural, although it may not be harmful. The life of Europe's mountains has been stamped by the mark of civilization, for better or worse. And to be sure, it has not always been for the worse. Man may have driven the alpine ibex, for example, to the brink of extinction, but man also brought this wild symbol of the European heights back from the brink.

Below; 92 top. *Known as the symbol of the healing arts, the Aesculapian snake* (Elaphe longissima) *has a vast range, from south-central Europe to Iran. In some parts of its range— the Alps, for instance—it reaches rather high altitudes, but it loves warmth and usually stays on slopes with a southerly exposure. The Aesculapian snake has a patchy distribution in central Europe. Some scientists suggest that such widely scattered populations may have been introduced by the Romans, who kept the snakes in the temples of Aesculapius, god of healing.* 92 bottom. *Native only to Europe, the asp viper* (Vipera aspis) *often ranges to considerable heights. It has been seen in the Pyrenees at an altitude of more than 2,000 meters. It is venomous but usually its bite is not fatal to man. It generally eats small mammals.* 94-95. *The European whip snake* (Coluber viridiflavus), *which may reach about two meters in length, is a good climber and can be found on the rocky slopes of the Alps and Pyrenees. It preys on small mammals, insects, reptiles, and amphibians.*

Africa
1. Rif
2. Atlas Mountains
3. Ahaggar
4. Tibesti
5. Semien Mountains
6. Ruwenzori Range
7. Aberdare Range
8. Drakensberg Mountains
9. Madagascar Plateau

The Heights of Africa

Africa is basically an immense, uplifted plateau of very rigid rock, dented with basins and edged with escarpments overlooking the ribbon of coastal plain around the continent. Rising from the underlying rock shield of the plateau are scattered mountain masses. They include peaks higher than the Alps and permanently capped with snow despite their location near the equator, volcanoes rumbling with activity, and the escarpments themselves, some of which drop precipitously from dizzying heights. The escarpments of the Drakensberg in Natal, Union of South Africa, and the Semien Mountains in Ethiopia, for example, abruptly fall more than 1,000 meters from their rocky rims.

Unlike the Alps, Caucasus, and Rocky Mountains, the mountains standing on the African shield show little evidence of folding (warping in bedrock caused by a movement of the Earth's crust). The rock that underlies Africa is so rigid that only in two places has the Earth been warped by folding into significant mountain ranges, and both are beyond the escarpments bounding the plateau.

These two regions—the Atlas Mountains of Morocco and Algeria, and the ranges at the southern tip of Africa— are separated by the full length of the continent. In between, folded mountains are virtually nonexistent.

As a rule, African mountains are the products of vulcanism or of faulting (cracking in bedrock involving a displacement of rock mass on one side in relation to rock on the other side)—sometimes both. Repeated outpourings of lava have piled up in many places to make plateaus created by mighty crags. At the top of the Drakensberg, for example, lies a cap of lava 1,000 meters thick. The most impressive example of mountains built by vulcanism, however, are in the continent's eastern equatorial region, where a flat landscape is dotted with towering, picturesque volcanoes, some with fires still seething within.

Several of the East African volcanoes—dormant Mt. Elgon on the Kenya–Uganda border, for instance—bulged from the Earth during the convulsions that marked the formation of the Great Rift Valley, the series of huge splits in the crust running southward from Ethiopia. Some of the other mountain masses were raised by the faulting which created the rift. The sheer heights of the Mau Escarpment in Kenya and the Livingstone Mountains in Tanzania, for example, are the elevated edges of the rift. The non-volcanic Ruwenzori Range, a rainswept rampart dividing Uganda and Zaïre, is a fault block heaved up at an angle from below the surface of the continent.

Isolated Ranges and Lonely Peaks

However they were formed, the mountains of Africa powerfully dramatize how isolation and stratification affect the nature of mountain life. Characteristically, African mountains stand in remarkable isolation from one another, separated by vast expanses of savannah, or desert, or in some cases by the clefts which form the Great Rift Valley. Mt. Kilimanjaro, the volcano in northern Tanzania which is the pinnacle of the continent, typifies this isolation. Kilimanjaro rises to an altitude of almost 5,900 meters, rearing its snow-capped summit above the surrounding flat savannah. The only other major mountain in the vicinity, Mt. Meru, is separated from Kilimanjaro's four peaks by almost 80 kilometers of grassland.

98-99. *The Great Rift Valley of Africa, extending southward from Ethiopia, is walled by towering scarps and contains a chain of large lakes, many of which are saline and some of which are among the deepest in the world.*

Above. *A toad* (Bufo hoechstii) *shows how effective natural camouflage can be as it hides against the granite rock of the Brandenberg Mountains in South West Africa.*
Right. *The flower* Variegata orbea *is one of approximately 10,000 species of flowering plants growing in the Cape region of South Africa. This region has a mediterranean climate, a type found elsewhere in Africa only at the opposite end of the continent, on the seaward side of the Atlas Mountains.*

For pure isolation, however, no mountains in Africa surpass those of Ethiopia. Rising to more than 4,500 meters above sea level, these mountains not only are far apart, but also stand atop an immense plateau which is 2,000 meters above the land around it. The Ethiopian plateau region, cut by deep gorges, encompasses most of the country; it is mitten-shaped, with the fingertips touching the Red Sea coastal region and the thumb jutting toward the horn of East Africa. Few mountain regions of the world are so spectacular and yet so bleak. The towering peaks and great chasms are the work of nature, while the stark, naked appearance of the rocky, desolate landscape has its origins in the activities of man. Once clad in thick, rich soil supporting lush vegetation, the Ethiopian highlands have been stripped of their cover so that vast expanses of highly productive land have been rendered useless. The pressures on the Ethiopian highlands have been the same as in other mountain regions of the world—poor agricultural practices and the need for timber and firewood by increasing populations. Nowhere, however, have the pressures been more severe. So seriously have native trees been depleted that the nation's capital, Addis Ababa, depends for lumber and fuel on the eucalyptus that has been introduced from Australia.

The animal life of the Ethiopian highlands, consequently, is but a shadow of what it was centuries ago. The creatures that remain, however, reveal that in ecological terms the mountains of Ethiopia are a point of contact between Europe and Africa, inhabited not only by a unique group of animals but also by creatures with affinities to the animals of Europe, as well as others linked to the fauna of equatorial Africa. The European chough (*Pyrrhocorax pyrrhocorax*), for example, shares the Semien Mountains with the purely Ethiopian Abyssinian wolf (*Canis simensis*) and the augur buzzard (*Buteo rufofuscus*) of East Africa. Ibexes—the Nubian (*Capra ibex nubiana*) and very rare Walia (*Capra ibex walie*) types—have made Ethiopian crags the southernmost outpost of the true goats, a Eurasian family; the mountain nyala (*Tragelaphus buxtoni*), which competes for food with domestic stock in the southern part of the country, represents a family of antelope, including the eland (*Taurotragus oryx*) and kudu (*Tragelaphus strepsiceros*), common in eastern and southern Africa.

On the Northwest Rim

In the northwest of the continent, the Atlas Mountains and the Rif Hills constitute the same type of zoological melting pot. Salamanders, characteristic of Europe but absent from the rest of Africa, inhabit the Rif, and the Atlas slopes are inhabited both by European species such as the wild boar (*Sus scrofa*) and by African animals such as the ratel (*Mellivora capensis*). In aspect, the Rif and the northern ranges of the Atlas seem European rather than African. On upper slopes, which intercept cold fronts and storms moving across the Mediterranean, snow falls heavily in the winter, mantling pines, cedars and similar conifers in white.

The mountains of Morocco and Algeria rise from a region with a climate that is typically Mediterranean, dry and subtropical. Like the folded nature of its mountains, northwest Africa shares its climatic type with no other part

Top. *Gelada baboons* (Theropithecus gelada) *range in troops through the Semien Mountains of Ethiopia. When attacked, these baboons may toss rocks down upon their enemies. During the night they take shelter in caves and under ledges.*
Bottom. *De Brazza's monkey* (Cercopithecus neglectus) *inhabits forests on Mt. Elgon and nearby mountains in western Kenya, as well as woodlands at lower altitudes in central Africa. This animal, one of the most brightly colored of all mammals, is a good climber but spends much of its time on the ground searching for insect prey.*

101

Above. *The Barbary sheep* (Ammotragus lervia), *which has goatlike traits, inhabits the Atlas Range and other mountains of northern Africa. It is a creature of dry rocky terrain.*
Right. *The foothills of the Atlas Mountains in Algeria have been denuded by excessive grazing, followed by erosion.*

of the continent except the Cape of Good Hope region. The latter is famous for its spectacular array of flowers—perhaps 25,000 different species—and is the native land of the gladiolus, and the rare disa orchid (*Orchis*), widely cultivated but rare in the wild. The orchid exists naturally only on Table Mountain, which overlooks Capetown.

Another trait that the mountains of the northwest rim share with those on the southern tip of Africa is ecological damage from agriculture and animal husbandry. In northwestern Africa, however, the environmental degradation is significantly worse. For 1,600 years the native Berber tribes with their sheep and goats have been increasingly confined to the mountains, first by the conquering Arabs, later by French colonists, and now by government farms on the fertile aprons below the heights. The concentration of tribal herds on the upper slopes has turned parts of the Atlas Range, for instance, into the same denuded and eroded landscape as the Ethiopian highlands. The disappearance of vegetation on such mountains makes them poor examples of stratification of life zones as related to altitude.

In Equatorial East Africa

Below Ethiopia, in the Cameroons, Guinea, and East Africa, mountains remain with large stretches of wilderness intact, although the lands surrounding many of them have been converted from jungle or savannah to farms and ranches.

But it is on the mountains of equatorial East Africa, particularly, that examples of stratification at its most striking can be seen. Nowhere else in the world is the range of environmental conditions from base to summit more extreme than on Kilimanjaro, Mt. Kenya, and the Ruwenzori. In few other parts of the globe does such a breathtaking collection of plants and animals populate mountain life zones. For instance, the slopes of the Ruwenzori are graced by a profusion of colossal heaths (*Erica*), far taller than a man, by giant lobelias (*Lobelia*) rising like spires in the mist, by gorillas (*Gorilla gorilla beringei*), leopards (*Panthera pardus*), and sunbirds (*Nectarinia*) of vivid red and green.

Allowing for some individual differences, the life zones of the mountains of this region follow much the same pattern as those in other parts of Africa, such as Ethiopia. A good example of such zones is found in the Aberdare Range, a particularly beautiful group of mountains in Kenya, on the western side of the Great Rift Valley.

The lower slopes of the Aberdares are clad in thick rain forest, the home of tree ferns (*Cyathea*) that raise their light green fronds towards the canopy. Here and there, particularly at lower levels, the forest is broken by glades of emerald-colored grass. In such open spaces buffalo gather, often in herds of scores of animals. These herds of the Aberdares seem to be a mixed lot. Some of the buffalo are large, with dark hides and heavy horns. Others are much smaller in build and horn, and reddish in color. In between the big dark ones and small red variety are buffalo of many shades and sizes. But all belong to the same species, *Syncerus caffer*. The variation in forms is probably due to a mixing of the two main races of the species, the savannah race and the dwarf forest type.

The savannah race is massive and dark, while the forest race is reddish and, like the forest varieties of many other mammals, comparatively small. Today the main range of the dwarf buffalo is in the vast forests of central and western Africa, but the smaller buffalo also has persisted where patches of forest remain among the savannah to the east. This may be the case in the Aberdares, an island of forest in the savannah, or of savannah converted to farmland. The Aberdares appear to be an area where the forest buffalo and the savannah race have met and integrated. Or else, the Aberdare buffalo may represent a third race, an entirely separate type suited to an environment intermediate between forest and savannah.

Grazing with the buffalo in the Aberdare clearings are countless wart hogs (*Phacochoerus aethiopicus*). Family groups trot about, tasseled tails stiffly erect, stopping to feed on juicy shoots, which the hogs select, like epicures, from among the grasses. Here and there the grass has been worn away, leaving circular depressions in the red earth, signs that the wart hogs have wallowed in such spots.

In some parts of the forest on the lower slopes, the earth is scarred by depressions much larger and deeper than the wallows of the wart hogs. These pits are craters left by bombs dropped by aircraft on the Aberdares when the mountains were a stronghold of the Mau Mau guerillas. The bombing, which occurred during the 1950s, is thought by some old Kenya hands to explain why elephants (*Loxodonta africana cyclotis*) in the neighborhood seem unusually edgy and often truculent. It is entirely possible that many of the pachyderms which still range the mountains were there when the bombs were dropped.

As the altitude increases, the forest grows on slopes that are almost sheer, cut by the white water of cascades and rising from deep ravines where bushbuck (*Tragelaphus scriptus*) feed. Bushbuck live all across the midsection of Africa and down its southeastern tier. They thrive in all sorts of environments, from flatlands to mountaintops up to 4,000 meters, but require either thick bush or trees. About the size of a roe deer (*Capreolus capreolus*), the bushbuck will often rest in the shade of a large shrub or small tree in the heat of the day, almost invisible in the shadows.

The canopy of the forest hides much of the life in it, but at times silvery-cheeked hornbills (*Bycanistes brevis*) can be seen flying under the green roof above, and black-and-white colobus monkeys (*Colobus polykomos*) emerge into the open from the tangle to munch leaves with great gusto. In the mountain forests, temperatures are somewhat cooler than in the surrounding lowlands, and the coat of the mountain colobus is even thicker and longer than that of their kin in the lowland jungles—an example of how variations within a species can result from differences as slight as a few degrees of temperature.

Approaching the 2,500-meter level, the Aberdare rain forest disappears, ending in a bamboo (*Arundinaria*) zone, a sequence followed on most mountains of tropical Africa. An exception is the quintessential African mountain, Kilimanjaro, on whose very dry upper slopes the rain forest dwindles to stunted scrub, resembling dwarf growth at the tree line of peaks in temperate regions. Generally, however, the pattern is one of rain forest fading into the

Above. *High in the Aberdare Mountains of Kenya, a troop of black-and-white colobus monkeys* (Colobus polykomos) *feeds in the thick bamboo groves. The colobus is related to the langur* (Presbytis) *of Asia, which also roams high in the mountains.*

106 top. *The eastern double-collared sunbird* (Cinnyris mediocris) *inhabits mountains from southern to eastern Africa and is usually seen at elevations above 1,500 meters. A bird of forest and scrubland, it also turns up in the backyard gardens.*
106 bottom. *The variable sunbird* (Cinnyris venustus) *flits among flowers in mountains from Eritrea south to Zambia. Nectar-feeders, sunbirds are the counterparts of the New World hummingbirds. Several species of sunbirds live high up in the African mountains, often above the timberline.*

109. *The sight of a bongo* (Taurotragus eurycerus) *is rare in the
bamboo forest of the Aberdares.
One of the most secretive and
mysterious of all antelopes, the
bongo moves through the bamboo
like a spirit, its harness of stripes
helping it blend into the foliage
and shadows. If cornered, it can
be a formidable foe.*
Above. *A broader view of the
bamboo forest is shown.*

bamboo, with trees such as juniper (*Juniperus*) on drier
sites, and yellowwood (*Podocarpus*), a member of the yew
family, signaling that the boundary is near. Mt. Kenya is
famous for these particularly remarkable stands of yellowwood and juniper trees that flourish slightly below the
bamboo zone.

Bamboo and Bongo

In the Aberdares, the transition from rain forest to bamboo
is abrupt. Trees are gone and are replaced by a nearly
impenetrable wall of grasses, some 20 meters high, for each
bamboo is actually a stalk of giant grass. The bamboo
forest is the home of the bongo antelope (*Taurotragus
eurycerus*), a handsome creature which moves through the
thick growth with uncanny ease and the elusiveness of a
phantom. The chestnut hide of the bongo, yoked with white
stripes, permits the 225 kilograms antelope to blend into the
shadows and foliage, where it browses on shoots and leaves,
although it probably does not eat bamboo. It sometimes
stands on its hind legs to reach leaves on higher branches.
Armed with powerful, lyre-shaped horns, the bongo often
is characterized as fearless; if compelled, it will stand its
ground against a leopard. Hunters who stalk the bongo
in the thick bamboo must be prepared for a sudden charge,
although generally a bongo that is disturbed will simply
retire into the shadows.

The bongo is not specifically a creature of the mountains,
but rather of the forest, either of the bamboo type or the
jungle. Only in East Africa, particularly in Kenya, is it
exclusively a mountain animal, simply because the only
forests in that region which suit it are on the slopes. Some
zoologists believe that the ancestor of the modern bongo
was a savannah species which retreated into deep woods. In
areas seldom disturbed by humans, in fact, the bongo sometimes roams just within the forest edges. Such places are
rare, however, so whether in lowlands or mountains, the
bongo keeps to the trees.

The capacity of the bongo to adapt to a wide range of altitude results from the tolerance of mammals for a variety of
temperatures. Like the bongo, many of the mammals roam-
ing the world's mountains below the timberline are not
especially adapted to mountain life but can simply get along
equally well in the highlands or lowlands. Usually, however,
they must be able to cope with forested conditions, for these
exist on the majority of mountainsides, or did before man
began to denude the mountains.

Except in the highest zones, the mammals of the East
African mountains belong, in general, to two groups. One
consists of a few creatures such as leopards and elephants
which can adjust to forest as well as bush and open plains.
The other group is made up of forest animals such as the
bongo and colobus. They are so specialized for the forest
habitat that they cannot live outside it, but they flourish
virtually wherever they find it, whatever the altitude. Life
in the uplands may cause some variations in a species—such
as the increased luxuriance of the coat in the mountain
colobus—but generally the mammals of Africa's mountain
forests also can be found in the forested lowlands. Gorillas,
for example, inhabit both the lowlands (*Gorilla gorilla
gorilla*) and mountains (*G. gorilla beringei*) of West Africa
and the mountains that divide East Africa from Central
Africa.

Gorillas Below, Orioles Above

Gorillas collect in groups ranging from a few to 30 animals, led by an older male. Contrary to stories about the savage nature of gorillas, they are giants of a gentle sort, exclusively herbivorous in diet—unlike chimpanzees (*Pan troglodytes*) and baboons—and generally placid. Although a gorilla when threatened is such a fierce adversary that a healthy adult gorilla has no true predators, the huge apes prefer to bluff an intruder rather than fight. The legendary chest-beating and screaming of the males is a means of cowing an opponent, rather than a form of attack. In fact, a gorilla will usually attack a human only if the chest-pounding routine frightens the man so much that he flees. This is why, in the Cameroons, people bitten by gorillas are considered cowards.

Moving on all fours, with the weight distributed on the soles of the feet and on the knuckles, gorillas range far up into the African heights, even well above the tree line. Scores of plants furnish the fruits, bark, and leaves which the huge apes eat, sometimes holding their meal with one hand while standing on their other three limbs. Gorillas pick their food by hand, then tear it with hands and teeth. As long as moist plants are available, gorillas do not drink water. The eastern race of gorilla, which inhabits the Virunga Volcanoes and other mountains where Zaïre, Uganda and Rwanda meet, often ranges far up, well above the tree line, a realm of mist and chill temperatures.

Birds, more sensitive to temperature, present a different picture in the mountains. The species in mountain forests contrast sharply with those of the lowland jungles. A few, such as the scaly francolin (*Francolinus squamatus*), manage to exist in both regions, but they are a small minority. More typical are the black-winged (*Oriolus nigripennis*) and green-headed (*O. chlorocephalus*) orioles which sing high in the canopy of the East African mountain forests but are absent from the woods of the lowlands. At the same time, birds inhabiting mountains that are separated by hundreds, even thousands of kilometers of African landscape tend to be alike, although they are not found in the intervening lowlands.

To many zoologists, the distribution of creatures such as the bongo, gorilla, and many mountain birds suggests that cool, moist forests once covered vast stretches of lowland between mountains that now stand in savannah or even desert. According to this theory, the forest expanded during cool, wet periods that corresponded to the Pleistocene glaciations in the northern continents. During interglacials, the forest presumably receded.

Oxford scholar R. E. Moreau, who studied the distribution of African birds, suggested a few years ago that a mountain forest environment linked highlands in West and East Africa, and Ethiopia and southern Africa, during the final glaciation—the Würm—which began 40,000 years ago. The forest expanded to its maximum, according to Moreau, about 18,000 years ago. Since then, the forest has retreated, leaving only relics, and their animal populations, on the East African mountains.

The mountains of East Africa form an archipelago of forest in which, just as on islands, individual species of animals have varied considerably, and closely related species have multiplied. For example, Moreau found ten different examples of the small birds called white-eyes (Zosteropidae)

110-113. *Mountain gorillas*
(Gorilla gorilla) *are sociable
creatures, living in bands which
on some mountains of east-central
Africa wander well above the
timberline. Like monkeys, gorillas
often groom one another. Although they seldom climb very
high in trees, they do scramble
up into the branches to feed and
sometimes make nests aloft to
spend the night.*

The rock hyrax (Procavia johnstoni) *has established itself in the upper zones of Mt. Kenya, between 3,200 meters and 4,650 meters. Very tolerant of cold, it can withstand temperatures down to minus 5°C.*

on separate mountains within a distance of 1,000 kilometers. On Mt. Elgon, a forested peak surrounded by near-desert, there are 26 species of carabid beetle belonging to a genus that lives mostly in northern Eurasia. The beetles probably all evolved from a very few ancestral types. The forests of the Mau Escarpment, on the northwestern flank of the Great Rift Valley, is the home of Peter's duiker (*Cephalophus callipygus*), a small antelope. The Aberdares, on the other flank, are inhabited by the related red duiker (*C. natalensis*), but the two species, separated by rift and savannah, are seldom in contact.

The Hyraxes

Among the most successful colonizers of many mountains in Africa, especially in the alpine regions above the bamboo zones, are curious little creatures called hyraxes (Procaviidae). Their distribution tells much about how animals from the lowlands establish themselves on the heights. Looking rather like woodchucks (*Marmota monax*) or guinea pigs (*Cavia porcellus*), but not related to either, hyraxes are lightning-quick inhabitants of rock piles in open areas, although one group, the tree hyraxes (*Dendrohyrax*), also has adapted to arboreal life in the forest. Highly versatile, hyraxes can eat almost any type of vegetation, and thus can survive even on the tough scrub of the alpine zone.

Hyraxes can be seen scurrying about the rocks of the alpine belts of most large mountains in tropical Africa. But the species differ according to the mountains, or more precisely, according to the nature of the surrounding lowlands. North of Mt. Kenya, for instance, lies a hot, arid expanse of savannah. From there have come rock hyraxes (*Procavia johnstoni*) that have colonized the heights above the bamboo. The same zone on the Ruwenzori Range, which is surrounded mostly by forest, is host to the tree hyrax. Further east, on Mt. Elgon, the alpine belt is the home of the yellow-spotted hyrax (*Heterohyrax brucei*). Below, on Mt. Elgon's forested slopes, live tree hyraxes. This suggests to some zoologists that the yellow-spotted hyrax, or its ancestors, appeared in the Mt. Elgon region before the tree hyrax and moved up the mountain before its arboreal relative arrived. From the evidence, it appears as if the tree hyrax moved in from West Africa late in the Pleistocene era, when forest linked the mountains of the east with the western part of the continent for the last time. The tree hyraxes then began to move up the slopes, but because they could not compete with the yellow-spotted hyraxes already established higher up, were held to the forest zone.

The High Moors

The alpine areas of African mountains above the bamboo zone resemble what one imagines the world was like in the Pleistocene era. Again, the Aberdares supply a vivid example of this world in the clouds. Above the bamboo belt, starting at about 3,000 meters, stretch vast, rolling moorlands sprinkled with the white flowers of everlasting and of St.-John's-wort (*Hypericum*) scrub. Icy streams cut across the moors, over whose rolling landscape clouds of mist are whisked by brisk winds. Here and there are tussock grass bogs, and now and then, very far off, foaming cataracts can be seen tumbling into hidden ravines. Scattered about the moors are remarkable thickets of giant heath, a relative of the small type that grows on the moors of Scotland, but of tree size. In the heath one sometimes can see elephants feeding, their huge, gray bodies moving through the gnarled, lichen-draped branches, white tusks gleaming against the dark foliage. The elephants trek up to the moors over muddy trails trodden deep into the sides of ravines so steep it seems incredible that these huge animals can traverse them.

The Pleistocene look of the moors is enhanced by the black rhinoceros (*Diceros bicornis*) and elands, other immigrants from the savannah below. The montagne francolin (*Francolinus psilolaemus*), a grouse-like bird, which seldom strays below the bamboo edge, is hunted on the moors by the serval (*Felis serval*), which pursues the birds with acrobatic leaps.

Almost 4,000 meters high, the Aberdares lack the snowcap of mountains such as Kilimanjaro, Mt. Kenya, and the peaks of the Ruwenzoris. On such mountains, there is a distinct and extensive "Afro-Alpine" zone above the heath, an eerie landscape where giant lobelias and senecios (*Senecio*) stand like ancient monoliths in the mist. Immense patches of these plants grow high in the Ruwenzoris, where tree hyraxes feed on them. Beyond this belt, the only forms of life to been seen are lichens and mosses, which carpet the bare rock. And farther up, the ice and snow begin, the frozen crown of Africa.

Increasingly, the mountains of equatorial East Africa are being hemmed in by expanding human populations and the farms and ranches necessary to feed those populations. Shambas—small farms—ring the Ngorongoro Crater of Tanzania. At the edge of the Aberdare National Park, which includes most of the range, a trench has been dug to separate the farms and their livestock from the creatures of the forest. Standing in the park, one can look out over wart hogs and buffalo while beyond, across the hidden trench, can be seen farmsteads and their cattle.

Mountains such as the Aberdares are no longer simply islands of forests among the savannahs—they are also the last of wilderness in a welter of humanity.

Top. *The black-headed oriole* (Oriolus larvatus) *ranges through much of tropical Africa, including the mountainous regions.*
Bottom. *A thick drapery of mosses festoons giant heaths* (Erica) *that grow at higher elevations in moist East African mountains. The heaths, taller than a man, are related to the knee-high heaths on the moors of Scotland.*
116-117. *Elephants* (Loxodonta africana) *climb high into the mountains of East Africa to feed amid mossy groves of giant heath growing on foggy, windswept moorlands just below the alpine zone.*

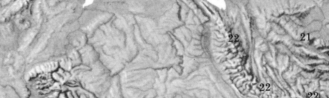

Asia

1. Taurus Mountains
2. Lebanon Mountains
3. Arabian Mountains
4. Jebel Akhdar Range
5. Alborz Mountains
6. Pamirs
7. Karakoram Range
8. Himalayas
9. Eastern and Western Ghats
10. Annamese Cordillera
11. Kun Lun Mountains
12. Tsinling Mountains
13. Nan Shan
14. Tien Shan
15. Altai Mountains
16. Sayan Mountains
17. Stanovoi Range
18. Sikhote-Alin Range
19. Japanese Alps
20. Sredinny Range
21. Cherskogo Range
22. Verkhoyansk Range

118

Asia and the South Pacific

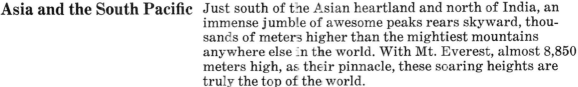

Just south of the Asian heartland and north of India, an immense jumble of awesome peaks rears skyward, thousands of meters higher than the mightiest mountains anywhere else in the world. With Mt. Everest, almost 8,850 meters high, as their pinnacle, these soaring heights are truly the top of the world.

The center of this mountain system on the northern fringes of the Indian subcontinent is the Himalayas, thrust toward the clouds from the bottom of a long-vanished sea by upheavals which began 65 million years ago. Starting near Pakistan in the west, the Himalayas form a jagged wall, rising out of steaming jungle and crowned by glaciers, while the eastern extremity grades into the Naga Hills and other highlands overlooking Burma.

The Himalayas, however, form only part of this enormous mountain system, the southernmost of several immense groups of mountains in mainland Asia. The western wing of the system arcs toward Europe, while the eastern wing reaches into China and also branches through the jungled hills of Indochina and the islands of the East Indies toward Australia. Like the principal mountain ranges of Europe, the Himalayas and those adjoining them have been a barrier between animals and plants adapted to different climates. During the glacial periods of the past, species which could not tolerate cold retreated south of the mountain wall, if they could manage to cross it. Today the endless expanses stretching north of the barrier to the hinterlands of Asia are largely cool and dry, while to the south conditions are largely warm and wet. The southeastern arm of the system gradually loses significance as a barrier as it loses altitude and finally merges with lowland jungle.

Mountains Beyond Mountains

The Himalayan center of the system described above is the most impressive mountain region on earth. In addition to Mt. Everest, the Himalayas have several peaks that are among the dozen highest mountains in the world. Mt. Kanchenjunga, in Nepal and Sikkim, is 8,598 meters high. Mt. Makalu, in Nepal and Tibet, is 8,481 meters above sea level and Mt. Dhaulagiri in Nepal stands 8,167 meters in altitude. Many other Himalayan peaks are almost as high.

West and slightly north of the Himalayas lies a wing of ranges with peaks that approach the Himalayas in height. The closest of these ranges to the Himalayas in distance as well as altitude is the Karakoram Range of Jammu and Kashmir, India, and extreme northern Pakistan. It contains the world's second highest mountain, K2, or Godwin Austen, which is less than 200 meters lower than Mt. Everest. In the Karakoram Range also are Mt. Gasherbrum, 8,068 meters high, and Mt. Rakaposhi, 7,790 meters high.

Further west is Hindu Kush, topped by 7,699-meter-high Tirich Mir. The Hindu Kush, in Pakistan and Afghanistan, is flanked on the south by the Hindu Raj, a smaller range but nevertheless one with peaks up to 6,400 meters in altitude.

To the east of the Himalayas, in China's Szechwan and Yunnan provinces, are several small but very rugged ranges. Notable among them is the Tahsueh Shan, through the valleys of which flow the upper Yangtze and Yalung rivers. These are the southernmost of central Asia's major moun-

Australia, New Zealand, New Guinea
1. Maoke Range
2. Owen Stanley Range
3. Great Dividing Range
4. Hammersley Range
5. Macdonnell Ranges
6. Musgrave Ranges
7. Snowy Mountains
8. New Zealand Alps

tain systems; north of them are range after range of
similarly imposing mountains separated in many cases by
high plains or plateaus. Until the latter decades of the
nineteenth century Western scientists believed that the
mountain ranges of central Asia were arranged in a lattice-
like fashion, but exploration of the region in the 1880s
proved them wrong. It was found that beginning with the
Himalayas, a series of parallel ranges and chains of ranges
continues all the way north to the southern margins of
Siberia.

In the Soviet Union immediately north of the Hindu Kush,
for example, is the Pamir-Allay range, with 7,482-meter-
high Mt. Communism. Further north, rising over the cold,
clear highland lake of Issyk Kul, are the foothills of the
Tien Shan range, which builds to altitudes of more than
7,000 meters to the east, in China.

The same parallel structure can be seen north of the
Himalayas. Their northern slopes merge with the southern
edge of the Tibetan Plateau. At the northern boundary of the
plateau rises another range, which like the Himalayas
runs generally in an east-west direction. This range is
the Kun Lun, over which stands 7,546-meter-high Ulugh
Muztagh. The next range in the series, north of the Kun Lun,
is the Altyn Tagh, beyond which lies the immense Takla-
makan Desert of Sinkiang. The northern edge of the desert
is rimmed by yet another range, the Chinese portion of
the Tien Shan, dominated by 7,439-meter-high Pobeda
Peak, on the Soviet border. North of the Tien Shan is the
arid tableland of Dzungaria, which in turn meets still
another range, the Altai, anchored to the west on the
Siberian border.

Similarly, the mountains east of the Himalayas form
parallel groups. North of the Tahsueh Shan is the Bayan
Kara Shan, then after the vast swamps of Tsaidam, the
Nan Shan, north of which is the Alashan Desert, then the
eastern extension of the Altai.

Besides these great mountain systems Asia has many other
ranges such as those of Siberia's Kamchatka Peninsula,
site of Klyuchevskaya Sopka, a volcano 4,749 meters high,
and the Sikhote-Alin range, a great wall rising above the
Sea of Japan north of Vladivostok. Western Asia has the
Zagros and Alborz mountains of Iran, more than 4,000
meters and 5,500 meters high, respectively.

A Profusion of Wildlife

The environmental conditions of this vast area run the
gamut of those found on the land. There is thus a tremendous variety of plants and animals in the region. In the
Sikhote-Alin range, wild ginseng (*Panax*) grows in isolated
patches hidden within the deciduous forests of the lower
slopes. On the high and stony barrens of the Tibetan
Plateau grows a low, clumpy plant known as *Saussurea*, distinguished by a fuzz covering which insulates it and holds
in moisture, and by a hole in its apex, through which bees
enter with pollen to fertilize the plant. There are rare
wild camels (*Camelus bactrianus*) on the slopes of the
Altai Mountains in summer, and giant pandas (*Ailuropoda
melanoleuca*) in the mountains of south-central China.
While the different habitats within Asian mountains
encourage a variety in species, there is also sufficient
similarity between mountain life zones in different ranges
to allow some species to be very widespread. Rhododen-

Himalayan langurs (Presbytis
entellus), *large monkeys related
to the colobus monkey of Africa's
mountains, inhabit rhododendron
thickets on the upper slopes in
Nepal and adjacent areas. Some
of the largest rhododendrons,
such as* Rhododendron arboreum,
are the size of small trees.

121

drons grow profusely at the edge of flower-speckled alpine meadows and moors high in mountains across Asia. Tigers (*Panthera tigris*) prowl the hills of Thailand as well as the mountains of Manchuria, Korea, and Siberia. Wild sheep and goats, closely related, inhabit the crags almost from one end of the continent to the other.

The Gorgeous Pheasants

Pheasants constitute a group of birds that is spread through several of Asia's mountainous regions, particularly in the central and southern part of the continent, and on some adjacent islands, such as those of Japan. Among the most beautiful of the birds—and all are stunning—are the monals, or impeyan pheasants (*Lophophorus*). The male monals have a thick plumage that shimmers with iridescent highlights of blues, greens, yellows, and oranges.

There are three species of monals. The Himalayan (*L. impeyanus*), the national bird of Nepal, is a chunky creature that perches on rocky spires and hops over the barren ground above the timberline up to 5,000 meters. When frightened, it resorts to a tactic that takes advantage of the high ground; it flaps heavily into the air, not far above the ground, and then sails downhill, a blue, red, and gold blur. Its speed is astonishing, and it can cover hundreds of meters in a few moments. But once it lands the monal must walk to return up the slope, because its wings are not strong enough for it to fly back.

The other monals live at somewhat lower altitudes. The Chinese monal (*L. lhuysii*), which inhabits western Szechwan, forages as far up as the timberline, and seldom climbs higher. Occasionally, when the weather is favorable, it may venture into the alpine meadows. The largest of the monals, it does not scratch with its feet for food as do some pheasants, but probes for food with its long beak. It spends most of its time on the ground but when night falls roosts either in rhododendrons or conifers. The plumage of the cock has a metallic gleam, which is intense even for a monal. Sclater's monal (*L. sclateri*), which ranges from the eastern Himalayas and southern Tibet to Yunnan, generally stays below 3,000 meters, where, like the Chinese species, it lives in the rhododendron thickets and among pines and firs.

Within these same rhododendron thickets lives another beautiful pheasant known as Temmick's tragopan (*Tragopan temmincki*). The tragopan cock is distinguished by the fleshy "horns" on its head and its bright blue lappet, or throat pouch, which it inflates during courting. Temmink's tragopan inhabits steep, rocky slopes from Tibet to Burma. The blood pheasant (*Ithaginis cruentus*), which has a throat of vivid scarlet, shares most of the range of Temmink's tragopan, and is notable because it seems to link the pheasants with the partridges. It is shaped much like a partridge, but like the pheasants it shows a signal difference in plumage between male and female. Female pheasants are for the most part drab, brown creatures, and lack the long, showy plumes of the cocks. The blood pheasant demonstrates the hardiness of the pheasants of the Asian mountains. Even in winter it does not descend below 2,000 meters, and in the summer it goes beyond the tree line, where its long whistle echoes over the meadows.

All but one species of pheasant are native only to Asia. The Asian highlands, especially, are the cradle of these gorgeous birds. Some, such as the Himalayan monal (Lophophorus impeyanus), *122 top, and the blue-eared pheasant* (Crossoptilon auritus), *above, are among the most spectacular of all birds. The silver pheasant* (Lophura nycthemera), *122 bottom, mentioned or depicted quite frequently in ancient Chinese literature and art, is known for its boldness. It has been widely introduced in Europe.*

124-125. *A male blood pheasant* (Ithaginis cruentus) *forages in the snow of the Himalayas, in Nepal. Except during the nesting season, these pheasants gather in flocks of sometimes more than a dozen birds. Thirteen races of this species inhabit the Asian mountains. In breeding season they form pairs, which may be monogamous. The cock and the hen contrast so much in color—the hen is brown—that they were once thought to be different species.*

Among the other high-country pheasants of mainland Asia are the brown-eared (*Crossoptilon mantchuricum*), which has immense white plumes; the Nepal koklass (*Pucrasia macrolopha nepalensis*), which is monogamous, a rare trait for a pheasant; and the spectacular Lady Amherst's pheasant (*Chrysolophus amherstiae*), which has huge, arching tail feathers, and a brilliant red crest, and is splashed with more than a half-dozen different hues.

The High Fliers
Myriad small but hardy Asian birds inhabit levels as high as or even higher than those of the pheasants. Several of these birds are closely related to kinds which live in the heights of Europe, and represent groups that have established themselves all across Eurasia in highland locations but separated by immense stretches of totally different lowland environments. The crossbill (*Loxia curvirostra himalayensis*) of the Himalayas, for example, is merely another race of the same species that twitters in the conifers of the Alps and other European ranges. The Asian variety never lives far from pine trees, in which it builds its nest of twigs and moss, and which provide the seeds that are its main diet.

In the Himalayas and nearby ranges there is a species of accentor that builds a loosely-constructed nest on the ground or in low vegetation, just as there is in the European mountains. The Asian maroon-backed accentor (*Prunella himalayana*) is much more colorful than its European relative, the alpine accentor (*P. collaris*). As the name of the Asian bird implies, its upper parts are richly colored, in hues ranging from maroon to chestnut. The neck and head are slate-colored, and the forehead is dappled with silvery patterns.

The rhododendron thickets growing below the alpine zone of the Himalayas and southeastern margins of the Tibetan Plateau are the home of such small birds as the fire-tailed myzornis (*Myzornis pyrrhoura*) and Gould's shortwing (*Brachypteryx stellatus*). Both feed on the hordes of insects that swarm through the thickets. The myzornis, green with a red tail, flits above the thickets, and specializes in picking off small insects that have been attracted to the rhododendron blossoms. The shortwing, with a chestnut back and vest of dappled slate gray, is more secretive, and hunts insects in the shady, damp inner recesses of the thickets.

Up the slope from the rhododendrons lives the swift, acrobatic grandala (*Grandala coelicolor*). The male grandala, deep blue with black wings, is a beauty, whereas the female, gray-brown streaked with white, is rather drab. Berry-eaters, grandalas nest at the edge of the snow zone, and leave the alpine level only when driven out by very frigid temperatures and heavy snow.

Except when they pair off during the breeding season, grandalas gather in large flocks sometimes numbering hundreds of birds each. A flock often will fly round and round, then land as a body on whatever vegetation is available. The long, pointed wings of the grandala enable it to fly powerfully, even when buffeted by the fierce winds that sweep across the heights of the Himalayas and the barren Tibetan Plateau.

Almost the entire expanse of the plateau is the nesting

126. *Lady Amherst's pheasant
(Chysolophus amherstiae)*
*another of Asia's spectacular
pheasants, has tail feathers about
a meter long.*
Above. *A tree nester, the satyr
tragopan (Tragopan satyra)
is one of the five species of
tragopans, all of which are
brilliantly colored. Tragopans
sometimes take over the aban-
doned nests of other birds, after
lining the old nests with leaves
and twigs. The usual clutch of a
tragopan is six eggs.*
Left. *Favoring dry, mountainous
country, the chukar partridge
(Alectoris graeca) ranges from
southern Europe to China, and
has been introduced as a game
bird in the arid mountain regions
of southwestern North America.
The bird gets its name from one
of its calls, which sounds like
"chuck-chuck-chuck."*

ground of the Tibetan snow finch (*Montifringilla adamsii*), which is very similar to the snow finch of the Alps and the Pyrenees. This bird feeds on seeds of the scant, low vegetation that manages to survive on the plateau, which has a growing season measured in weeks rather than months. Various types of edelweiss (*Leontopodium*) flourish on the plateau, demonstrating again that it is often altitude rather than geographical location which determines the distribution of mountain species around the world.

Waterfowl of the Plateau
Parts of the plateau are dotted with lakes and marshes which are the homes or migration rest stops for waterfowl and other water birds such as plovers and sandpipers. Breeding in these wetlands and lakes is the bar-headed goose (*Anser indicus*), which gets its name from a pair of black markings on the back of its white head. When winter approaches, the goose migrates southward, flying so high it passes over the crest of the Himalayas. This remarkable bird has even been seen winging through the crystal sky over Mt. Everest.

Home of the Pandas
The extreme northeastern edge of the Tibetan Plateau, largely a region of steppes near the Bayan Kara Shan, sometimes is visited by the giant panda (*Ailuropoda melanoleuca*), a creature whose stronghold is slightly east of the plateau, among the great ravines and sheer mountain cliffs of western Szechwan. The region is a mass of mountains and valleys copiously watered by fog, rain, and snow. Some of the valleys, shielded from the cold winds and winter storms by mountains to the west, are parklike, covered with grass in summer, and dotted with poplars, oaks, and yews. For the most part, however, the land is stark, jagged, and altogether awesome. Through this spectacular landscape surge the upper waters of great rivers—the Mekong, Yangtze, Hwang Ho, and Yalung.
On the mountainsides, beginning at about 1,800 meters, is the zone of bamboo that is the primary home of the giant panda. This bear-size beast roves to the upper edges of the bamboo, which generally grows to about 2,500 meters but on some slopes extends several hundred meters higher, where it mixes with stands of silver fir. Known to Western science for little more than a century, the giant panda has given taxonomists matter for endless debate about its position among the families of mammals. It has been shifted back and forth between the bear family and the one to which the raccoon (*Procyon lotor*) belongs. It looks rather like a bear, particularly because of the similar bulk, but there is another species of panda which looks like a large, orange-red raccoon, even to the face mask and ringed tail. This is the lesser panda (*Ailurus fulgens*), which is about a meter long from snout to tail and shares part of its range with its large, black-and-white cousin.
The lesser panda inhabits a much broader area of the Asian mountains and also customarily ranges to higher altitudes than the giant species. An agile climber, the lesser panda can be seen above the bamboo, in forests of pine and fir, and amidst thickets of huge rhododendrons, gnarled and hung with mosses and lichens, and standing in thick mats of sphagnum moss. The rhododendrons form

128. *Insects have colonized the heights of mountains. Many insects are able to get along well above the timberline even in the glacial snow. One of the reasons insects can survive very high in the mountains is that they are able to resist freezing far more than most other creatures. Mainly because of a compound called glycerol, the body of the insect does not freeze even at temperatures below 0°C. Insects have even been found in Antarctica.* Top, *orb weaver* (Argiopidae) ; *center, short-horned grasshopper* (Acrididae) ; *bottom, dragonfly* (Sympetrum).

Left. *Many beautiful flowers bloom on the middle slopes of the Himalayas of Nepal. The* Porona grandiflora, *top, grows at about 2,400 meters. The* Caltha palustris L., *center, and* Geranium doniannum, *bottom, were photographed at 3,000 meters. Insects are attracted to the flowers on the slopes, and in turn support populations of small birds.*

130-131. *Ancestor of almost all domestic geese, the graylag goose* (Anser anser) *of eastern Europe and Asia has been seen flying high above Asian peaks on its trans-Himalayan migrations. This goose is a creature of open country, although it does not nest away from water.*

a nearly impenetrable tangle, which in early summer is ablaze with the colors of their blooms. The flowers speckle the mountain slopes with reds, pinks, yellows, and whites, brilliant amidst the glossy green leaves of the shrubs.

Often the giant panda joins its smaller relative in the rhododendron thickets, and sometimes even climbs higher, more than 5,000 meters above sea level, to roam among the poppy-dotted grasslands just below the snowfields. There the panda digs for bulbs, although its main diet consists of bamboo stalks and shoots, harvested below the rhododendron zone. Despite its large size and strength, the giant panda is not really a predator, although it may on rare occasions eat bird's eggs, insects, a mouse or similar small animal.

Bears and Wolves

Considerably more carnivorous are the two species of bears which also inhabit the mountain home of the giant panda. From Tibet, a race of the brown bear (*Ursus arctos*) sometimes wanders into panda country. Brown bears, as noted earlier, are native to much of Eurasia, and although they have been exterminated from large portions of their range, still hold out in most of the mountain regions from Turkey and the Urals east to Japan, and from the Himalayas through Siberia.

The other bear likely to be encountered in the land of the giant panda is a species unique to Asia, *Selenarctos thibetanus*. Although sometimes called the Himalayan black bear, it inhabits a much more extensive area than the name implies. The range of this black bear extends from Iran to the northern islands of Japan, and within this vast region the animal lives mainly in mountainous areas. The species was believed to be extinct in Iran until 1973, when biologists discovered that several black bears still roam the Makran Range of the southeastern province of Baluchistan and the adjoining province of Kerman. The bears seem to prefer forests of wild pistachio at lower elevations and date groves at higher levels.

Both black and brown bears, particularly the former, have a reputation for killing domestic livestock. Although some kills undoubtedly occur, it may also be that the bears frequently scavenge the bodies of dead sheep and cattle, and are wrongly blamed for the killings. Both species of bears eat a substantial amount of animal matter, although their diet also includes plants of many types.

Like bears, wolves (*Canis lupus*) inhabit much of Asia north of the tropics, and are found in several mountain regions. In highlands such as the Pontic Mountains of eastern Turkey, wolves are numerous enough to approach villages and even big towns in large packs when driven out of the heights by winter storms. In places such as the Hindu Kush, wolf packs regularly raid domestic herds in hit-and-run style.

The Big Cats

Three of the big cats are among the other large predators found in several of the scattered Asian mountain regions. One, the Siberian tiger (*Panthera tigris altaica*), is the largest cat in the world. It weighs more than 300 kilograms, and can reach almost four meters in length, including the tail. Paler, with a much heavier coat than tigers of southern

132. *Giant pandas* (Ailuropoda melanoleuca) *seem to be almost always hungry. These large creatures of the mountain forests spend up to half of the daylight hours eating. Like the lesser panda, the giant species sometimes feeds on small mammals, and it also occasionally catches fish, but the major part of its diet is plants, mostly bamboo.* Top. *The lesser panda* (Ailurus fulgens) *is a mild-mannered creature, largely arboreal in habit. It is more active by night than by day.* Bottom. *The Asiatic black bear* (Selenarctos thibetanus) *is a large predator found high in the mountains of central Asia. During the summer the bear ranges up to 3,600 meters.*

*The heavy pelage of the
snow leopard* (Uncia uncia or
Panthera uncia) *accounts for its
ability to live on the cold heights
of central Asia. Man has eliminated the snow leopard from
most of its range and it is
threatened with extinction.*

climes, the Siberian tiger once lived through most of northern China, Korea, Manchuria, and Siberia. Today, zoologists believe the huge cat has healthy populations only in a few mountainous areas—the Sikhote-Alin of the Soviet Union, and in the mountains on either side of the Yalu and Tumen rivers in an area centering on the Chiang Pai Range.

Two, and perhaps three, other races of tigers are also found in some of the Asian mountains, although they are by no means true highland animals. The Indochinese race has found refuge in some of the jungle-covered hills of its native region. There are still tigers, for example, in hilly Khao Yai National Park, only two hours' drive by automobile from Bangkok. In northern India and Nepal, tigers of the Bengal variety survive in the foothills of the Himalayas. A few Caspian tigers may still roam the mountains on the border of Iran and Afghanistan, but many zoologists consider the race extinct.

Like the tiger, the leopard (*Panthera pardus*) is a cat that includes mountains among its habitats. The leopard, however, is much more widespread, and immensely more numerous than the tiger. No other big cat seems nearly as adaptable as the leopard, which remains in many regions from which the tiger has been exterminated. They survive in nearly all the large mountain regions of Asia.

The third big cat of the Asian mountains, unlike its relatives, is truly a predator of the heights, living nowhere else. It is the snow leopard (*Uncia uncia*), and is among the most beautiful of cats. It is cream-colored on its chest and underparts, with a thick coat of ghostly gray, marked by black rings or rosettes, above. Few people have ever seen this creature, for it shuns men and never leaves the high places, where its coat blends in with the rock and snow. Uncommon everywhere, snow leopards nevertheless have a wide range, from the Hindu Kush east through Tibet and the Himalayas to Szechwan, and northeast through the Pamir-Allay, Tien Shan, Altai, and the Sayan Mountains, a small range on the border of Mongolia and the Soviet Union.

High in these mountains, the snow leopard hunts for large prey, chiefly the wild goats and sheep that thrive on the Asian peaks. From the very few observations that have been made of the snow leopard, it seems that it may hunt in the early morning and late afternoon, but so seldom has the cat been studied that it is impossible to be sure of this. It is certain, however, that the continued survival of the snow leopard is linked to the survival of the wild goats and sheep of the mountains.

Wild Sheep

Many of the wild sheep of Asia have been greatly diminished by the destruction of their highland habitat and by excessive trophy hunting. Two species have been particularly targeted by trophy hunters because of the size of the animals' horns and bodies. The argali (*Ovis ammon*) of the high plateaus from Russian Turkestan to Tibet is the largest wild sheep, with a body the size of a large donkey and horns that spiral for more than 150 centimeters across its front. Closely related to the argali is the Marco Polo sheep (*O. ammon polii*) of the Pamir-Allay, Hindu Kush, and Karakoram mountains. It is slightly smaller than the argali, but has even larger horns.

136-137. *In parts of its range, the tiger (Panthera tigris) roams in the mountains: Bengal tigers reach the foothills of the Himalayas. The Indochinese tiger has retreated to the mountainous areas such as those around Khao Yai National Park in Thailand. The Siberian tiger inhabits mountains in northeastern Asia—the Sikhote-Alin in the Soviet Union and mountains on either side of the Yalu and Tumen rivers.*

All of the many true wild sheep are closely related, and indeed, most zoologists believe that nearly all of them are simply variations of the same species. In western Asia, for example, live several sheep that are similar in all purposes except for different shades of color or size.

The urial sheep (*O. ammon arkal*), which stands a meter high at the shoulder, weighs 85 kilograms, and has a white ruff, inhabits the rolling foothills east of Iran's Alborz Range, a barrier just south of the Caspian sea. In the Alborz lives the Alborz red sheep (*O. ammon orientalis*), smaller, and darker than the urial of the east, and is distinguished by a black chest marking below its white ruff. The Alborz sheep is an intermediate form between the urial and the Armenian urial (*O. ammon gmelini*) of northwestern Iran. The Armenian urial weighs only 20 kilograms, has no white ruff, but has a large black chest patch. In southwestern Iran, chiefly in the Zagros Mountains and neighboring highlands of Fars and Kerman, lives yet another wild sheep, the Larestan sheep (*O. ammon laristanica*), almost as small as the Armenian urial and with an extensive black patch on chest and neck.

Wild Goats

The highest peaks of the Alborz and Zagros mountains are the home of another type of horned, hoofed mammal, the Persian wild goat (*Capra aegagrus*). Of stocky build, with short legs characteristic of many mammals living in cold places, the wild goat is the ancestor of domestic goats. The wild goat lives in the mountains of Turkey and Iraq as well as Iran, and a few of them survive in the White Mountains of Crete.

The bearded, powerful male wild goat, which can weigh 90 kilograms, carries huge, curving horns, which sometimes exceed 140 centimeters in length. The female is smaller, with short slender horns. Mating takes place in the autumn and twin kids are born in early spring, after the goats have spent the winter in snow-free areas on the lower slopes.

Closely related to the Persian wild goats are the ibexes, several of which are found in Asia. Small ibexes of the Nubian race populate Sinai and the rocky mountains of the Negev and Judaean Hills of Israel. The large Siberian ibex (*Capra ibex sibirica*) inhabits the mountains of the Altai, Tien Shan, the Palmir-Allay, and nearby ranges. There, during the summer, it feeds amidst the sedum (*Sedum*), vetches (*Vicia*), and louseworts of the high meadows, and even climbs to the snow line. Only a decade or so ago, the Siberian ibex was not considered to be in peril, but now it, too, has suffered a dangerous population decline. In winter, the ibex descends to lower altitudes to feed and take shelter. Increasingly, however, a wave of human settlement has swept up the slopes of the Asian mountains, pushing ever higher in search of firewood and forage for livestock. Caught between the rigors of the wintry heights and the crush of the people below, the ibex is running out of space. At the same time, it, too, has suffered from overhunting. The markhor goat (*Capra falconeri*), for instance, has been eliminated from vast stretches of territory where until recently it was numerous, and even common.

Famous for its huge horns, which form tight spirals, this marvelous creature was one of the first to feel the effects

Left. *The markhor* (Capra falconeri) *is known for its spiral horns and the dense fringes of hair on its body. Heavy snows drive the markhor out of the heights in winter, but at other times this big wild goat will range well above the timberline.* Above. *Notable for its heavy mane, the Himalayan tahr* (Hemitragus jemlahicus) *is becoming increasingly scarce in its native Asian mountains. The tahr is a taxonomical puzzle, some scientists considering it a goat, others putting it in a group of its own.*

Tough and courageous, the wild yak (Bos grunniens mutus) roves the high desert steppes to Tibet, a region that lacks trees. The yak must keep on the move to find enough forage. The wild yak is much larger and more aggressive than the domestic variety, which is used as a beast of burden and for its milk.

of human activity in the mountains, because in many regions it inhabits somewhat lower altitudes than the other goats and sheep. As early as the 1930s, zoologists reported that the markhor was being killed off by gun-carrying tribesmen in the mountains of Afghanistan, a prime part of its range. Today the situation is even worse. The only place in which it gets effective protection is in the Soviet Union, where perhaps 1,000 markhors live.

Tahr, Blue Sheep, and Antelope

The Nilgiri tahr (*Hemitragus hylocrius*), a goat-like creature that lives at altitudes up to 1,800 meters in the hills of southern India, has suffered even more severely. Less than a thousand Nilgiri tahr survive on the rolling plateaus of their homeland. The closely related Arabian tahr (*H. jayakari*), which inhabits the barren dry Jebel Arhdar range of Oman, is similarly imperiled. The Himalayan tahr (*H. jemlahicus*) has fared better.

The "blue sheep," or bharal (*Pseudois nayaur*), has been rather removed from such pressures because it dwells in unusually remote regions of the mountains of central Asia. It is not really a sheep but an animal with the traits of both sheep and goats. The center of the bharal's range is the steppes of the Tibetan Plateau, where the animal can be found on heights almost up to 6,000 meters. It grazes on grassy slopes in herds that may number several hundred animals. When frightened, this heavily muscled animal bolts for the nearest cliffside, where it takes refuge. This type of behavior is more typical of the goats than the wild sheep of Asia, which generally rely on outrunning their enemies.

Running over the steppes of the Tibetan Plateau are two small antelopes, creatures not generally associated with high mountains. One is the Tibetan gazelle (*Procapra picticaudata*), which inhabits the most southerly and less rigorous part of the plateau. The other is the Tibetan antelope, or chiru (*Pantholops hodgsoni*), a strange, little-known animal related to the bulbous-nosed saiga antelope (*Saiga tatarica*) of the Eurasian grasslands. The small, delicate chiru, which seldom weighs more than 30 kilograms, inhabits one of the most demanding regions on earth, the steppes of northern Tibet, a tableland generally above 5,000 meters in altitude, barren of trees and shrubs, and almost of grass. Virtually the only vegetation in significant amounts occurs around scattered marshes and small lakes.

The Wild Yak

This bleak region is also the home of the wild yak (*Bos grunniens mutus*), an awesome creature that in behavior and stature is immensely more imposing than the placid domestic yak (*Bos grunniens grunniens*). A thousand kilograms in weight, with curving horns whose tips can be as far apart as 90 centimeters, it is the monarch of its mountains. Unlike the small domestic yak, which has been turned into a riding animal as well as a beast of burden, the wild variety brooks no interference from man. Bulls are as formidable as any African buffalo (*Syncerus caffer*), and if provoked by humans will erupt into a furious charge, bent on destruction. Shaggy, with a thick winter coat, it spends the summers in the snowfields, sometimes above 6,000 meters. In winter, the yak herds move to

Serows (Capricornis suma-traensis) *are here seen on Mt. Shiga in Japan. This curious beast, related to the chamois, inhabits the lower slopes of wet mountain ranges in Japan and Southeast Asia. The serow is a powerful animal and when pressed by man or beast will defend itself with its dagger-like horns. Serows seldom leave heavy bush and tend to use the same runways on a regular basis.*

marshes and lakes, and into the valleys. In the face of
blizzards, which can swirl over the plateau even in early
summer, the yaks, which generally get their moisture
from eating snow rather than drinking water, turn their
rumps into the storm and wait for it to subside.

Goat Antelopes

The goat antelopes, which include the chamois of Europe
(*Rupicapra rupicapra*) and the Rocky Mountain goat of
North America (*Oreamnos americanus*), are also represented in the mountains of Asia. None of the three Asian
members of this group is alpine, but one, the takin
(*Budorcas taxicolor*) of southern China and northern
Burma, sometimes climbs to an altitude of more than 4,000
meters in the summer. Oxlike in build, with a stocky body
that may weigh up to 350 kilograms, the takin seems
clumsy, particularly because its front legs appear oversized.
But while the animal is not especially fleet, it easily makes
its way up and over rocky hillsides with movements that are
precise and deliberate, if not graceful.

Related to the takin are the goral and serow, which are
smaller than their relative but much more swift. Neither
ascends to the altitude that the takin does, but in summer
the goral (*Naemorhedus goral*), which inhabits the Himalayas and the mountains of Korea and Manchuria, may go
almost to an altitude of 4,000 meters. In winter, the gorals
try to avoid moving about in the snow because they become
bogged down in it and thus an easy prey for their enemies.
The serow (*Capricornius sumatraensis*), distinguished by its
heavy neck mane, lives on the lower, jungled slopes of the
Himalayas, south through Indochina to Sumatra and also
in Japan. In Indochina, serows sometimes come down to
sea level but generally remain up on steep limestone cliffs,
which because of their inaccessibility to man are still
covered with jungles. Such cliffs, which provide a haven
for many different animals, are pocked with caves; the
serow take shelter in these.

Mountain Creatures of Australasia

South of the Asian mainland, on the islands of the region
generally recognized as Australasia, are several mountain ranges with peaks that qualify them as major mountains. The highest are on the island of New Guinea, where
Mt. Carstensz stands just centimeters short of 5,000 meters
amidst the Nassau Mountains of West Irian. In the nearby
Orange Mountains is Mt. Wilhelmina, 4,750 meters high,
and in the Bismarck Range in the Papua, New Guinea,
portion of the island is Mt. Wilhelm, 4,693 meters above
sea level.

The mountains of New Guinea are the home of the little-
known marsupial, the forest mountain wallaby (*Dorcopsulus
macleani*), which roves at night in the misty jungles that
reach above 3,000 meters in altitude. These wallabies have
longer forelimbs and hind limbs that are not as powerful
as those of the larger kangaroos in open country, indicating
that they may scramble about the undergrowth rather than
travel by leaps and bounds.

In the same highland forests scientists have recently discovered an exceedingly strange type of ecological com-
munity. A group of minuscule animals—protozoa, rotifers,
lice, and mites—inhabit miniature forests of lichens, moss,
fungi, and liverworts growing on the backs of weevils

The nocturnal rock wallabies
(Petrogale penicillata), top, *and*
(Petrogale xanthropus), bottom,
*spend much of the day in caves
and rock shelters. These mar-
supials are master climbers and
can leap across chasms up to four
meters in width. In many places
inhabited by rock wallabies, the
stone of cliffs has been worn
shiny by the endless procession
of these animals.*

only two centimeters long. The moisture in the soggy
atmosphere of the mountain forests keeps the mini-plants
growing on holdfasts among the many tiny protuberances
and pits on the backs of the weevils, which themselves are
plant eaters. While carried about by the weevils, the animals
living in the micro-communities aboard the insects feed on
the tiny plants, or on one another, or on refuse, keeping
their tiny worlds in balance.

Several of the birds of paradise inhabit the high reaches
of New Guinea's mountains. The four sickle-billed species,
for example, seldom descend below 1,500 meters. The black
sickle-bill (*Epimachus fastuosus*) is the heaviest of all birds
of paradise, and approaches the size of a crow (*Corvus
brachyrynchos*), to which these beautiful birds are related.
With a metallic green glint to its coal-colored plumage, and
long fluttery shoulder plumes which are fanned during
courtship, the male black sickle-bill typifies the loveliness
of this fascinating group of birds.

Australia has many regions of steep, rugged terrain, such
as the Great Dividing Range that walls the moist east
coast from the dry interior, but no mountains of truly
great height. The highest mountain in the continent, Mt.
Kosciusko, in the Australian Alps, is less than 2,230 meters
above sea level. Many of the mammals of lowland Australia,
therefore, live in mountains as well. If there are any creatures in Australia that can be said to be specialized for
mountain living, they are the rock wallabies (*Petrogale*).
The soles of their feet are treaded pads surrounded by stiff
hairs— an arrangement that acts rather like the hoof of the
mountain goat mentioned earlier in promoting sure-footed-
ness on rocks. Unlike the typical kangaroo tail, moreover,
that of the rock wallaby is tufted at the tip, which enhances
its capacity as a balancing organ when the creature leaps
from ledge to ledge or among boulders.

New Zealand's Alpine Birds

New Zealand's Southern Alps, dominated by 3,764-meter-
high Mt. Cook, is a truly mountainous area which has
sheltered some unique birds. A thousand meters up in this
range lives an unusual flightless bird called the takahe
(*Notornis mantelli*), thought to be extinct until it was re-
discovered in 1948. Actually, the takahe has been resur-
rected from extinction twice. Scientists first knew of it only
from fossil remains, and it was presumed to belong to the
past. But in 1849 the living bird was discovered. Larger
than a rooster, the takahe is a gallinule, a bird similar to
the rails. According to where its remains have been found,
it once inhabited both of New Zealand's major islands.
Before the beginning of this century, the takahe was believed to be extinct, and occasional reports that it had been
seen or heard in the remoteness of the New Zealand Alps,
the wildest part of the nation, were discounted by most
scientists.

An ornithologist named G. B. Orbell believed the reports.
In 1948 he led an expedition into the region of the Alps
called Fjordland, because of its resemblance to the mountainous coast of Norway, and netted two takahes, which
he filmed and then released. Once Orbell's evidence was
presented, the government made the area inhabited by the
takahes a reserve, and the wild population is now carefully monitored.

Like other gallinules and rails, the purple-hued takahe

has long, spreading toes that enable it to skitter over lily pads and other aquatic vegetation, despite the bird's considerable size. In the wet, upland habitat, generally above the Antarctic beech forests of the lower mountainsides, the takahe forages for grasses, eating both the flowering tufts and the stems.

Sharing the Fjordland fastnesses with the takahe is an even rarer bird, the kakapo (*Strigops habroptilus*), or owl parrot, a strange, nocturnal creature that can glide but is virtually incapable of flying. Instead, the kakapo runs along the ground and creeps about in the trees. Thus handicapped, it was vulnerable to the dogs, cats, stoats, and other mammalian creatures introduced into New Zealand by humans. Introduced red deer (*Cervus elaphus*), while not the enemies of the kakapo, were competitors for its food, and consumed the leaves and berries of the wet beech forests inhabited by the big-eyed parrots-of-the-night. Like so many other animals elsewhere, the kakapo was forced up into the mountain refuges where its enemies and its competitors were not numerous. The kakapo now clings to existence in the heart of Fjordland, in a few patches of remote beech forest soaked by more than 700 centimeters of rain annually.

Above the beech forests, and the scrub-dotted moorlands inhabited by the takahe, stretches the alpine zone of the New Zealand Alps. It is the habitat of yet another strange bird that has suffered at the hands of man but has now begun to flourish once more. The bird is the kea (*Nestor notabilis*), and, like the kakapo, is a parrot unique to New Zealand. Bold, aggressive and adaptable, the kea searches the alpine meadows for food. Its diet consists of about 40 different plants, including gentians and daisies, as well as insects and other small animals. The kea and man met only in 1856, and almost immediately came into conflict, because the kea acquired a taste for sheep flesh. Ranchers began to find the bodies of dead sheep torn by the sharp bills of the parrots. Indeed, keas could sometimes be seen perched on the carcasses of sheep. Keas were killed on sight, and a bounty was offered for every kea beak presented to the government. Yet despite the persecution the kea survived in its high country home. Today, the bird's reputation as a sheep killer has faded, and zoologists believe that rather than attacking live sheep, the kea scavenges those that it already finds dead.

Understanding the true nature of the kea has encouraged toleration of this bird, whose toughness and adaptability typifies the qualities animals must possess to inhabit the high country.

Below. *A bold, inquisitive creature, the kea* (Nestor notabilis) *is one of the few parrots that must regularly cope with seasonal snowfall. It feeds on meat—often carrion of sheep—but in the summer consumes large quantities of nectar.*

146-147. *Still hardly explored, the mountains of New Guinea support lush rain forests. Several of the birds of paradise inhabit mountains such as the Bismarck Range of Papua, New Guinea.*

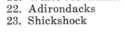

North America
1. Brooks Range
2. Alaska Range
3. Chugach Mountains
4. Wrangell Mountains
5. St. Elias Range
6. Coast Ranges
7. Cascades
8. Olympics
9. Sierras
10. Mackenzie Mountains
11. Rockies
12. Uintas
13. Wasatch Mountains
14. Eastern Sierra Madre
15. Western Sierra Madre
16. Sierra Madre del Sur
17. Sierra Madre
18. Talamanca Range
19. Smokies
20. Appalachians
21. White Mountains
22. Adirondacks
23. Shickshock

The Americas:
The Great Cordillera and
Other Ranges

The main mountain system of the Americas, in contrast
to chains such as the Alps, Pyrenees, and Himalayas, have a
distinct north-south orientation, and cut a swath through
many latitudes. Unlike the cordilleras of Eurasia, the one
that forms the spine of the American continents runs along
the length, rather than the breadth, of the landscape.
Beginning in Alaska near the Arctic Circle, it stretches in
an almost unbroken series of mountain chains along the
western margins of the Americas to the storm-wracked tip
of South America, not far above the Antarctic. In the
eastern United States, and to a lesser degree in eastern
Brazil, smaller mountain ranges show the same north-south
arrangement.

Because these ranges, particularly the great cordillera,
traverse so many latitudinal zones, an ever-changing parade
of plants and animals can be encountered along their peaks
and slopes. No other cordillera has a flora and fauna that
contrasts so markedly, not only up and down the mountainside but along its length, as that of North and South
America. At one place or another one can find creatures as
diverse as caribou (*Rangifer tarandus*), flamingoes
(Pheonicopteridae), dwarf boas (*Ungaliophis continentalis*), grizzly bears (*Ursus arctos horribilis*), jaguars
(*Panthera onca*), Rocky Mountain goats (*Oreamnos
americanus*), hoary marmots (*Marmota caligata*), and
guinea pigs (*Cavia porcellus*). The plants growing on the
heights of these great ranges run the gamut from reindeer
moss (*Cladonia*), arctic poppies (*Papaver*), Sitka spruce
(*Picea sitchensis*) and giant sequoias (*Sequoia gigantea*) to
orchids (*Orchis*), southern beech (*Nothofagus*), cacti
(Cactaceae), podocarps (Podocarpaceae), and palms
(Palmae).

Fingers from the North
Amidst such variety, however, are many plants and
animals ranging literally from one end to the other of
some of the major American mountain chains. The out-
standing ecological characteristic along the crests of these
mountains are fingers of northern vegetation that reach
thousands of kilometers south of the limits of similar plants
in the lowlands. Atop 3,861-meter-high San Francisco
Mountain in the Arizona Rockies, for instance, arctic
tundra like that of northern Canada and Alaska overlooks
the desert-like grassland of the American Southwest.
On the highest ridges and summits of the Appalachians,
in the southeastern United States, grow thick woodlands of
spruce and fir, which in the lowlands are not found further
south than northern New England and Canada.

Appalachian Wilderness
The ancient Appalachians, which begin as sea-lashed cliffs
in Newfoundland, stretch for more than 3,200 kilometers
to the southeast where, after rising to a height of more
than 2,000 meters, they fade into the lowlands of Georgia.
The Appalachian system is composed of myriad regional
and local ranges, which while they often appear to form
a tangled maze, are generally arranged in parallel series,
all tending toward the southeast. The White Mountains
of New Hampshire, for example, are made up of the
Presidential, Franconia, and a few lesser ranges. Flank-
ing the White Mountains on the west are the Green
Mountains of Vermont, which in turn are paralleled

150-151. *The awesome Columbia Glacier in the Chugach Range in Alaska drops its icebergs into the sea at Prince William Sound. Icebergs are carved from the glacier at high tide. In summer, with the exception of the glaciers, the range is snow free.*
Above. *With an immense range, from one end of the Americas to the other, the cougar* (Felis concolor) *is a highly adaptable cat. Long persecuted by man, the cougar has vanished from many regions. It was thought to be extinct in the Appalachians, but recently a small population was discovered there.*

further west, across Lake Champlain, by the primeval Adirondacks. Further south, the wall of the Blue Ridge is paralleled to the west by the magnificent peaks of the Great Smokies.

All in all, the Appalachians cross 17 degrees of latitude and cover more than 160,000 square kilometers of territory. Even though some of this region lies on the doorstep of the megalopolis that sprawls between Washington, D.C., and Boston, Massachusetts, it contains vast stretches of wilderness.

Although the Appalachians begin in Canada, they are not spectacular until they reach Maine. There, the rugged eminence of Mt. Katahdin, 1,605 meters high, rises above a damp, cold conifer forest, dotted with ponds and bogs, the home of the porcupine (*Erethizon dorsatum*), lynx (*Lynx lynx canadensis*), and white-tailed deer (*Odocoileus virginianus*). While not imposing in terms of altitude, Katahdin, crowned by crags and boulders, is impressive for its stark and primal bleakness. Indeed, a combination of location, weather patterns, and geology has given many mountains in the northern part of the New England states qualities that keep them truly savage and primeval wildernesses, like much higher ranges elsewhere.

Nowhere in the Appalachians is nature more untamed than in the White Mountains, specifically the lonely heights of the Presidential Range, which rises from a rolling country-side dotted with farms and towns and only 200 kilometers from Boston. The White Mountains are composed of granite, which once surged out of the Earth's depths in a molten state but instead of breaking through the crust pushed it into a blister. The "skin" covering the blister vanished long before man appeared even in the most primitive form, but the granite, hard and resistant, remained, although subjected to some of the most destructive forces that have ever scourged the surface of the Earth.

Boulders, of a type that could only have been carried from the north and dropped upon the heights of the White Mountains, show that they were entirely covered by the glaciers of the Pleistocene era. The scars left by the glaciers are almost everywhere in the White Mountains. Rock has been grooved and scratched by the passage of the huge ice sheets. Round valleys, called cirques by geologists and "gulfs" by New Englanders, have been gouged into the landscape. Cirques form when local glaciers expand, scooping out chunks of rock from the earth. There are no glaciers in the White Mountains today, but one could form in the most famous of cirques, Tuckerman's Ravine—where snow lasts until July—if there were a slight but sustained drop in summer temperatures.

Tuckerman's Ravine lies in the Presidentials, where eleven peaks tower more than 1,500 meters in height. Startlingly light in color, the Presidentials loom over their approaches like much higher mountains—an impression fostered by their rocky prominences and the fact that they stand almost 1,400 meters above their immediate surroundings. Ascending the slopes, one moves through forests of sugar maples (*Acer saccharum*) and beeches (*Fagus grandifolia*), then red spruce (*Picea rubens*) and balsam fir (*Abies balsamea*). The latter species continues up the mountainside and, along with black spruce (*Picea mariana*), persists to the timberline.

The timberline in the Presidential Range can occur at less

than 1,400 meters, for the range lies in the path of several major storm tracks and is continually raked by high winds. Along the more than 13 kilometers of tundra and alpine meadows that crest the Presidentials, snow may fall in any month of the year. Soft and soggy underfoot, the tundra is strewn with boulders of varying sizes, and during the warmer months is often cloaked by fog and mist. Almost continually, on summer days, the haunting call of the gray-cheeked thrush (*Catharus minimus*) can be heard ringing up from the ravines which drop precipitously into the conifers below. About the only birds which can be seen regularly on the tundra are the white-throated sparrow (*Zonotrichia albicollis*) and the junco (*Junco hyemalis*), and even they leave with the approach of winter. So rigorous is the environment that mammals are scarce and small. The red-backed mouse and a few shrews (*Sorex cinereus; Blarina brevicauda*) are virtually the only common types.

West of the White Mountains, the Appalachian ranges are not as stark. The Green Mountains of Vermont are gentle and rolling, while the Adirondacks of New York, although rugged, are lower and not nearly as bleak as the Presidentials. The Adirondacks are, however, venerable. Their foundation is part of the Canadian shield, shoved up from the continental basement over a billion years ago, and then much later incorporated into the Appalachians when the system was formed by more recent upheavals.

Return of the Cougar

Nowhere in the Appalachians are there any strictly large alpine mammals, such as the Rocky Mountain goat and chamois. The Appalachian forests, however, teem with deer of the white-tail or "Virginia" species. The deer herds have prospered as never before because they have been nearly free of predators other than feral dogs, and more recently, a few coyotes (*Canis latrans*) which have spread east from regions to the north and west; and because forests have been cleared to make fields which provide better foraging for deer. Wolves (*Canis lupus*) long ago were exterminated in the Appalachian region, and so, it was believed, was the cougar, or "mountain lion" (*Felis concolor*). But in the 1960s and early 1970s even skeptics agreed that the cougar had returned, as elsewhere, as a predator to the eastern mountains, albeit in small numbers.

The cougar, like the leopard (*Panthera pardus*) of the Old World, is a highly adaptable big cat, with a range that extends from British Columbia to Patagonia. While it is often thought of as a mountain beast, it also inhabits deserts, plains, jungles, and even wetlands, such as the Florida Everglades. But mountains have become the stronghold of the tawny cat, and it is still relatively common in parts of the Rockies and the Pacific Coast ranges of North America. In the Appalachians, the eastern race of the cougar was deemed extinct for many years—at least, that is what zoologists believed. From Canada to Georgia, reports of cougar sightings continued to occur, although rarely. Some of the reports were from areas remarkably close to urban centers, including New York City. Eventually, between sightings of cougars and finding of tracks and similar signs, enough evidence was accumulated to satisfy zoologists that the cougar did roam the Appalachian wilderness.

154-155. *The cougar obtains its prey by stalking, creeping close, and then bounding onto its victim. Cougars are a major predator of deer but also eat animals as small as rodents.*

Above. *Large numbers of American kestrels* (Falco sparverius), *small falcons, migrate south over the Appalachian ridges in October. Kestrels prey on insects and tiny mammals such as mice.*

157 top. *The bald eagle* (Haliaeetus leucocephalus) *can be found in mountains along the sea. This large bird, whose numbers have been depleted because of pesticide contamination, feeds largely on fish, which it catches, steals from ospreys and diving ducks, or finds dead on the beach.*
157 bottom. *A common scavenger in much of North America, the turkey vulture* (Cathartes aura) *lives year-round in the southern Appalachian region, but migrates south from the northern parts of the chain.*

Flights Over Hawk Mountain

A wildlife spectacle for which the northern Appalachians
are most famous is the autumn migration of birds of prey
which follow the mountain ridges southward. Migrating
eagles and vultures on their way south take advantage of
thermals—updrafts of warm air, which rise when winds
sheer against the sides of ridges. In the low mountains of
eastern Pennsylvania, several skyways come together and
funnel past a rocky lookout known as Hawk Mountain. On
fine autumn days, hawks, eagles, ospreys, and vultures
by the thousands glide past the lookout, which has the
reputation as the best site in North America, if not in the
world, to watch birds of prey in flight. More than a dozen
species of hawks and eagles regularly migrate past the
lookout, and on good days the procession of birds never
ceases.

August brings the first of the migrants, which as a rule are
led by a few bald eagles (*Haliaeetus leucocephalus*), followed
by squadrons of broad-winged hawks (*Buteo platypterus*),
chunky birds that are the type most commonly seen flying
past the lookout. By September, the broad-winged hawk
migration begins to peak, and more species join the flights
heading south. Small sharp-shinned hawks (*Accipiter
striatus*)—accipiters related to goshawks (*A. gentilis*)—
flit by. The large red-tailed hawks (*Buteo jamaicensis*)
have begun to appear, and by the end of October will make
up the major portion of the migrants. Now and then a
golden eagle (*Aquila chrysaetos*) brings shouts of admiration from the crowds of nature enthusiasts who gather atop
the lookout to watch for the birds. As the weeks pass,
Cooper's hawks (*Accipiter cooperii*), red-shouldered hawks
(*Buteo lineatus*), and even a few rare peregrines (*Falco
peregrinus*) wing past.

Hawk Mountain is in a region that contrasts sharply with
the Appalachians of New England. The mountains south
of New England are not jagged and sharp, like the Presidentials, but rather a series of long, almost unbroken
ridges. Their smooth outline is due to the fact that they
were never scourged by the ice sheets.

Blue Ridge and the Smokies

Toward their southern limits, the Appalachians stretch
into their most elongate ridges and reach their greatest
heights. The Blue Ridge, the first mountain wall to be
encountered by European settlers as they moved west from
the southeastern coastal plain, stretches for some 1,300
kilometers southeast from the Potomac River. To the west
of the Blue Ridge, across the picturesque Shenandoah
Valley, lie the rough and crumpled Alleghenies, and then,
further south, the Great Smokies, the heartland of the
Appalachians.

The Smokies get their name from the puffy clouds of
mist and fog that rise from their well-watered slopes and
valleys. The range has about two dozen peaks higher than
Mt. Washington, surmounted by the highest mountain east
of the Mississippi River, Mt. Mitchell, 2,037 meters high.
Atop such peaks, the weather resembles that of southern
Canada and northern New England. During the severe
winters, up to 90 centimeters of snow accumulates in the
heights of the Smokies, while 10 centimeters at most covers
the lowlands around the range. Spring comes several weeks
later to the high Smokies than to surrounding regions. By

May, the valleys of eastern Tennessee and western North Carolina are green, sun-warmed, and speckled with flowers, while up above, near the summits, roads and trails can be blocked with snowdrifts five feet deep.

Very likely the higher zones of the Smokies were covered with tundra during the ice ages, and the lowlands around them forested with the red spruce and Fraser fir (*Abies fraseri*) that now dominate the landscape above the 1,800-meter level. When the cold of the late Pleistocene era moderated, the lowlands became too warm for these conifers, and the spruce-fir forest retreated up the slopes until it was squeezed into the only area where it can now survive, atop the highest peaks. Today these conifer forests isolated in the heights of the Smokies are islands of the so-called Canadian life zone. While there is no true alpine zone in the Smokies, there are grassy meadows and heaths which resemble it. Some scientists believe that these open areas, called "balds," are the result of a warming trend soon after the end of the last ice age. During the warm period which followed the glacial times the climate was even more moderate than it is today. The spruce-fir forest could have been eliminated from some mountaintops by the warmth and then failed to recover when the climate turned cooler again. On a few such balds, however, the conifers seem to be making a slight comeback. The resulting combination of scattered conifers and the vegetation of the balds has created some of the most beautiful natural rock gardens anywhere. One of the finest is on Roan Mountain, which straddles the border between Tennessee and North Carolina.

Interspersed with the outcroppings of gray granite on the gentle slopes leading to the summit are groves of spruce, meadows of mountain oat grass (*Arrhenatherum*), and thickets of rhododendron (*Rhododendron carolinianum*). In spring and summer, the meadows are flecked with the colors of wildflowers—violets (*Viola*), the rare Gray's lily (*Lilium grayi*), red trillium (*Trillium erectum*), wild geraniums, and many more. In June the rhododendron thickets are ablaze with blood-red, purple, and white flowers. Edging some of the meadows are wild azalea bushes (*Rhododendron calendulaceum*), some of which have hybridized, and which add a mix of reds and oranges to the spectacle.

The astounding variety of habitats which exist in the Smokies makes them a cross-section of the Appalachians, and thus within their boundaries can be found most of the animals which live in this mountain system. The porcupine and lynx are absent, but the bobcat (*Lynx rufus*)—smaller relative of the lynx—can be heard yowling in the night. The American black bear (*Ursus americanus*) is common, and so is the ubiquitous raccoon (*Procyon lotor*). During the hours of darkness, flying squirrels (*Glaucomys volans; G. sabrinus*)—which do not really fly but glide by means of flaps of skin on their sides—zoom between the trees. Muskrats (*Ondatra zibethica*) and a few beaver (*Castor canadensis*) ply the waters of ponds and streams, and in some parts of the mountains can be seen a visitor introduced from the Old World—the European wild boar (*Sus scrofa*).

The birds at lower altitudes in the Smokies reflect the avian populations of adjacent lowlands. Cardinals (*Cardinalis cardinalis*) send their melodious calls

from the treetops. The Carolina wren (*Thryothorus ludovicianus*), song sparrow (*Melospiza melodia*), tufted titmouse (*Parus bicolor*), phoebe (*Sayornis phoebe*), and Carolina chickadee (*Parus carolinensis*) live there year-round. In the spring, vireos (Vireonidae), a variety of warblers (Parulidae), the gorgeous indigo bunting (*Passerina cyanea*), and the Acadian flycatcher (*Empidonax virescens*) appears. The screech owl (*Otus asio*) calls eerily in the night, and crows (*Corvus brachyrhynchos*) flap overhead.

Higher up, the bird life resembles that of northern New England. Ravens (*Corvus corax*), rather than crows, hover in the sky. The black-capped chickadee (*Parus atricapillus*) replaces the Carolina variety, the veery (*Catharus fuscescens*) takes the place of the wood thrush, and the saw-whet owl (*Aegolius acadicus*) is heard instead of the screech owl. Canada warblers twitter in the rhododendrons, and crossbills (*Loxia curvirostra*) perch in the spruces. Juncos (*Junco hyemalis*), gray birds of the north, flit here and there close to the ground, but leave for the lowlands during the most severe winter weather.

Scuttling around the lower slopes of the Smokies are lizards, which are typical of the southeastern United States and not found in the north. Carolina anoles (*Anolis carolinensis*) and brown skinks (*Lygosoma laterale*) are common below 800 meters. Higher up can be found the swift, or fence lizard (*Sceloporus undulatus*), which has a correspondingly more northerly range than the others. The venomous copperhead (*Agkistrodon contortrix mokasen*) and timber rattlesnake (*Crotalus horridus*), which range almost the length of the Appalachians, are fairly common in the Smokies. During the winter these snakes gather in large numbers within dens in rocky areas, and hibernate. The moist, woodland habitats of the Smokies and neighboring ranges in the southern Appalachians are ideal for salamanders, which thrive there. Almost thirty different salamanders inhabit the Smokies alone. Several of the salamanders which inhabit the southern parts of the Appalachian system live within extremely precise altitudinal parameters, and some are unique to the region, even to the very specific localities within the mountains. The red-cheeked salamander (*Plethodon jordani jordani*), for example, is found only within the Great Smoky Mountains National Park. The Roanoke salamander (*P. wehrlei dixi*), purple-brown with bronze mottling, lives only in caves within the mountains of southwestern Virginia. Metcalf's salamander (*P. jordani metcalfi*) inhabits the mossy floor of the Blue Ridge forests. The salamander is restricted to a very precise altitudinal range between 1,066 and 1,767 meters.

The Metcalf's, Roanoke, and red-cheeked salamanders are all closely related. Metcalf's and the red-cheeked in fact, are different versions of the same species. Metcalf's salamander is black on the back and light underneath. The red-cheeked has a similar body color but bright rosy patches on its cheeks. Another member of the same species, the red-legged salamander (*P. jordani shermani*), has a black body with bright red legs. All of these salamanders, and several other closely related forms, inhabit their own particular area of the southern Appalachians. Scientists believe that most, if not all, descended from a common ancestor that once was widespread throughout the mountains, but that

160; 162-163. *A moist, woodsy environment makes the southern Appalachians a haven for salamanders, which teem among the humus on the woodland floor near the small streams of the region and among caves and clefts in the rock. Seldom seen, the salamanders nevertheless abound. They are most often encountered in late winter and early spring, their breeding season. Then they migrate to breeding waters, often crossing highways in large numbers on rainy nights.* 160: top, *red-legged salamander* (Plethodon jordani shermani); center, *green salamander* (Aneides aeneus); bottom, *red-cheeked salamander* (Plethodon jordani jordani). 162: Row 1: left, *spotted salamander* (Ambystoma maculatum); center, *marbled salamander* (Ambystoma opacum); right, *cave salamander* (Eurycea lucifuga). Row 2: left, *northern dusky salamander* (Desmognathus fuscus fuscus); center, *pygmy salamander* (Desmognathus wrighti); right, *Yonahlossee salamander* (Plethodon yonahlossee). Row 3: left, *black-chinned red salamander* (Pseudotrian ruber schencki); center, *spring salamander* (Gyrinophilus porphyriticus); right, *red eft* (Diemictylus viridescens viridescens).

climatic changes, perhaps during the ice ages, isolated pockets of the original population in various mountain strongholds and, during the long ages of isolation, each population of salamander evolved somewhat differently. Out West, in North America's true high country, the bighorn sheep (*Ovis canadensis*) has undergone a similar differentiation, although its population was spread over a much wider area—almost half of the continent, in fact. The bighorn, originally ranging from northern Canada into Mexico and eastward to Nebraska, split into more than a half dozen different races, some of which have vanished in the last century and all of which are now rare. Today, bighorn populations are scattered and few. Still, the clash of horn upon horn as the rams battle for supremacy in the spring can be heard at one place or another in all three of the immense chains that, running parallel to one another, form the North American section of the great cordillera.

The Western Ramparts

The western ramparts of the continent begin at the very edge of the Pacific with the Coast Ranges, which in Alaska are backed up by the mighty Alaska Range, crowned by 6,190-meter-high Mt. McKinley, the tallest peak in North America. East of the Coast Ranges, south of British Columbia, it begins in the north with the Cascades, studded with volcanic peaks such as Mt. Rainier and Mt. Hood, and culminates in the Sierra Nevada, which stretch for more than 600 kilometers to the south until they sink into the rock and sand of the Mojave Desert.

In Canada, the Cascades–Sierra Nevada wall is squeezed between the Coast Ranges and the easternmost rampart, the Rockies, and lost in a welter of saw-toothed highlands. The Rockies, longest mountain chain on earth, begin in Arctic Alaska with the Brooks Range. At its highest, only about 2,800 meters, the Brooks Range bulges out of landscape that is already above timberline, not because of altitude but rather because of high latitude. Further south, the Rockies become a jumble of more than 60 smaller ranges, which together make up a system running all the way to Mexico, and to the edges of the American Great Plains.

The western slopes of the Rockies complex are generally more heavily forested than those on the eastern side, because by the time the winds from the Pacific have crossed the crest of the mountains, most of the moisture carried from the sea has been lost. As a rule, however, the entire Rockies chain is much drier than the mountains of eastern North America.

Bighorn!

There are in a sense two groups of flora and fauna in the western mountains. One is composed of species distributed throughout all or most of the three mountain chains. The other is made up of types that are peculiar to very small areas within this vast mountainous region. The bighorn sheep belongs in the first group, because before it began to suffer from competition with domestic livestock and from a disease called scabies, as well as from poorly managed hunting, it lived from the eastern slopes of the Cascades to the edges of the Great Plains. Moreover, if (as some zoologists believe) the thinhorn, or Dall, sheep is the same

A race of the brown bear (Ursus arctos) *especially noted for its ferocity, the grizzly bear* (U. arctos horribilis), *shown here, occasionally climbs into tree branches when young but as an adult generally remains on the ground. South of the Canadian border, the grizzly has become rare, except in a few mountain strongholds.*

species as the bighorn, then the bighorn's distribution extends north of the Arctic Circle, through the Brooks Range.

Throughout this region, bighorn sheep live among many different plant communities and altitudes, ranging from 4,000-meter-high meadows to the bleak depths of Death Valley, a sun-scorched graben that dips almost 90 meters below sea level. In the Sierras, bighorns winter in the lower parts of canyons, facing east, below 2,000 meters, among plants such as bitterbrush (*Purshia tridentata*), ricegrass (*Oryzopsis hymenoides*), pinyon pine (*Pinus monophylla*), and speargrass (*Stipa speciosa*). Bighorns of the desert mountains in southwestern Arizona roam a parched landscape dotted with cholla (*Opuntia ramossima*), teddy-bear cactus (*O. bigelovii*), and creosote bushes (*Larrea divaricata*). A group of bighorns that has summered atop Wyoming's Mt. Washburn, 3,108 meters high, feeds in alpine meadows covered with *Carex* sedges and dotted with saxifrage (*Saxifraga*), plants that root in crevices of rocks. While generally similar in conformation, the races of bighorn inhabiting these many regions differ in certain characteristics, notably in the size and shape of horns and in coloration.

During the winter, the bighorn depends on the shelter and forage available at lower altitudes, which are now occupied by ranches, farms, and ski resorts. The presence of man has forced the animals out of many of their wintering areas. Other than man and disease, the enemies of the bighorn are few. Coyotes occasionally take bighorn lambs and a few infirm adults. Golden eagles, rulers of the spaces over the mountains of the Americas as well as of Eurasia, sometimes swoop down on young lambs. Wolves once were a menace, but today south of Canada are all but gone. The cougar, which prowls most of the western mountains, remains dangerous for sheep that cannot gain the rocky heights in time to escape.

Great Bear of the West

Much of the same region originally inhabited by the bighorn was included in the range of the grizzly bear (*Ursus arctos horribilis*), one of the races of brown bear in North America. Weighing more than 200 kilograms, and sometimes twice that, the grizzly needs up to 270 square kilometers of home range per bear. While it usually does not go out of its way to attack people, it is terrifying when aroused; it will charge if startled by a human and in fact has claimed lives. Not surprisingly, when settlers poured into the western United States, they and the grizzly bear became mortal enemies. Today, grizzlies remain common in Alaska and Canada, but less than 1,000 survive below the Canadian border. These few bears are feeling increasing pressure as people pour into national parks and forests seeking to commune with nature, and as the mountain regions are increasingly exploited for energy, minerals, and timber. Although a grizzly will occasionally attack man, especially if the bear happens to be a sow with cubs, and will almost never pass up a wapiti (*Cervus elaphus*) or moose (*Alces alces*) mired in snow, it is not primarily a predator. The only creatures that grizzlies hunt with any regularity are small rodents such as ground squirrels (*Citellus*). For the most part, the great bears are peaceable eaters of grass, berries, and roots. They demand solitude, however, and

even in the wilderness of the mountainous west of North America, this is an increasingly rare condition.

Music of the Mountains

In the mountains all creatures must contend with deep snow. Some of them avoid it entirely. The yellow-rumped warbler (*Dendroica cononata*) for example, heads for the Pacific Coast or Mexico. The black-tailed prairie dog (*Cynomys ludovicianus*) retires to its burrow, although it does not truly hibernate. The woodland caribou (*Rangifer tarandus*) of the Canadian Rockies and the wapiti, which lives in the Rockies from British Columbia south and along the central Coast Ranges, descend to the valleys. In Yellowstone National Park, the wapiti, largest race of the red deer, leaves the timberline in November and travels substantial distances—sometimes scores of kilometers—to winter ranges down below.

With the coming of spring, the wapiti returns to the high country. By midsummer the great, branching antlers of the bulls have grown to their maximum, and by September these temporary weapons are ready for duels with other bulls over the right to mate with the females. This is the time that the mountains echo with the haunting sound of bugling, the musical challenge of the bulls.

The most magnificent of the many races of wapiti is the Roosevelt wapiti (*Cervus elaphus Roosevelti*), which inhabits the Coast Ranges from northern California to southern British Columbia. Large, dark in color, with ponderous antlers, this big deer has a stronghold in the wild Olympic Mountains of Washington State, near the Canadian border.

Not nearly as high as the Rockies, the Olympic Mountains are nevertheless among the most spectacular highlands in the Americas. They are the site of a true temperate rain forest, where Sitka spruce (*Picea sitchensis*) grow almost 100 meters tall and Douglas fir (*Pseudotsuga menziesii*) approach five meters in diameter. Soggy under as much as five meters of rain, mist, and snow a year, the forests of the Olympics are laden with mosses and are as dank and dark as any tropical jungle. The rainfall of this region is the heaviest in the continental United States, resulting from the proximity of the mountains to the sea. They are the wall upon which break the moisture-laden winds that have swept across the Pacific. Once the winds strike the mountains, they drop their moisture, and stay bone dry until more moisture is picked up on the way to the Rockies.

The Big Trees

The two westernmost mountain chains—the Coast Ranges and Sierra Nevada—are havens for three spectacular species which are survivors from prehistoric times. Two of these are members of the plant kingdom and are among the most spectacular of all plants. Growing on the western slopes of the Sierra Nevada at altitudes of between about 1,200 meters and 2,400 meters is the mightiest of all trees, the giant sequoia (*Sequoia gigantea*). It is rivaled for size by its relative, the coast redwood (*S. sempervirens*), which grows on slopes up to about 600 meters in altitude within a belt up to 48 kilometers wide on the mist-shrouded Coast Ranges of northern California. Before the ice ages, the coast redwood was widespread in North America and western Europe, and the giant sequoia grew over large areas of

western North America. Today, they survive only in the limited areas just described, and although the logging that in the last century destroyed thousands of giant sequoias has been curbed, the timber industry still threatens the last untouched stands of redwood.

It would be a calamity to lose either of these species of tree, because their sheer size makes them wonders of the natural world. Giant sequoias approach 90 meters in height and in the past may have been even taller. Based on the size of a well-known fallen trunk known as the "father of the forest," the tree of which it once was a part may have towered 121 meters above the ground. Around its base, a giant sequoia can measure more than 30 meters, and its diameter can be 10 meters or more. The redwood is a little less broad, but loftier than the existing sequoias. A redwood discovered in 1966, for instance, stands 117 meters high. Such growth takes enormous amounts of time. The largest sequoias began as tiny seeds the size of a pin-head, before the time of the Exodus. The oldest redwoods date from the time of Christ.

The Giant Birds
In the Coast Ranges a short distance north of Los Angeles, California, a narrow dirt track loops up the mountainside, around hair-raising curves, to a promontory overlooking a landscape of canyons and high ridges stretching to the horizon. If one waits, has sharp eyesight, and is lucky, a great bird may be seen, soaring on wings set straight out, with tips curved forward, and measuring more than three meters across. It is the California condor (*Gymnogyps californianus*), a huge vulture very similar to the type that during the Pleistocene era feasted on the carcasses of beasts that died in the famed La Brea tar pits, now the site of downtown Los Angeles.

The California condor numbers no more than a few dozen individuals and is close to extinction, despite the fact that two large sanctuaries have been set aside for it in the California mountains that are its last haven. Once quite common, the condor has been in decline for more than a century, ever since the American settlement of California. During the California Gold Rush and the years that followed, people shot the big birds for fun, or because they seemed to menace livestock, although in truth they eat only carrion. More recently, condors have perished from eating poisoned bait set out for coyotes and other predators. But what endangers the condor most of all are its own breeding habits. When nesting, which it accomplishes in clefts or crannies on rock ledges, the big but shy bird is extremely sensitive to disturbance and will abandon its single egg or young at almost any interference More-over, condors live for almost a half century but do not breed until they are six years old, and even then nest only on alternate years.

Frogs with Tails and Other Oddities
Because of the tremendous expanse of territory covered by the North American cordillera, many odd or rare animals are found within its boundaries. Some of these zoological curiosities are so unusual that they stand out, even among the bizarre. In rushing streams from the Cascades to the western Rockies, for instance, lives the tailed frog (*Ascaphus truei*). The name comes from a tail-like copulatory

Largest of the red deer (Cervus elaphus), the wapiti summers high in the Rocky Mountains and the Coast Ranges. Its favorite food is grass, but it has an extremely varied diet and can either graze or browse. The wapiti breeds in September-October, when bulls try to gather harems of as many cows as possible. Young are born the following spring. The calves stay with their mother throughout the summer and into the winter. In late winter or early spring the bulls shed their antlers. A new pair, however, begins to grow and by summer the bulls are in velvet.

organ carried by the male, which fertilizes the female
internally, a habit unique among frogs. This frog is also
unusual in that it seems voiceless. The chief function of
voice in the frogs is to attract mates, and it may be that in
the rushing waters inhabited by the tailed frog, a piping or
croaking love call might not be heard by the females.
Another cold-blooded oddity of the North American cordillera is the gila monster (*Heloderma suspectum*) of the
southwestern United States. One of the only two species of
venomous lizards in the world, this chunky, pink-and-black
creature sometimes ascends desert mountains to altitudes
above 1,000 meters.
In rivers of the Kern Plateau in the southern Sierra
Nevada, above 3,000 meters, is found a unique species
of trout, the golden trout (*Salmo agua-bonita*), which
since its widespread transplantation to other waters
has become most beloved of anglers. The golden trout
probably developed from a population of rainbow trout
isolated when streams were cut off after the ice ages.
There it developed the distinctive golden and yellow highlights that give this beautiful fish its name.

Cloud Forests of Latin America
Isolation on a mountain has created a unique species of
dwarf boa (*Ungaliophis continentalis*) in the Sierra de
Juárez of northeastern Mexico. The boa, a black reptile less
than a half meter long, was discovered beneath a rock in a
cloud forest in 1967. The cloud forest is a wet habitat of the
tropics and subtropics, and forms on mountainsides which
are cloaked much of the time in fog and mist. A warm-
climate counterpart to soggy slopes in North America's
Coast Ranges, the cloud forest of Latin America is generally
found on slopes that are rather protected from the winds
and only midway up mountainsides. Exposed, mountaintop
cloud forests dwindle into the stunted, "elfin woodland," like
that atop the Luquillo Mountains of Puerto Rico, described
earlier in these pages.
From Mexico almost to the glacier-capped hook at the southern end of South America, cloud forests can be found all
along the great cordillera. While these watery woodlands
exist elsewhere in the world, they are most extensive on the
American cordillera.
Under assault by hordes of impoverished farmers who
practice slash-and-burn agriculture, and subject to the
same sort of developmental pressures as wild lands in the
northern reaches of the American cordillera, the cloud
forests are dwindling. This is a tragedy, for with their high
humidity, cool nights, and hot days, these forests are extremely rich in plant life and contain countless species
which have never been identified. Some of these plants may
be valuable for their medicinal properties. Others, such as
orchids, certainly should be treasured for their fabulous
beauty.
More than 1,000 species of orchids grow in the cloud forests
of Central America alone. Many are air plants—that is,
they do not root in the soil but grow on perches in the forks
of branches or in crevices on the trunks of trees. Some of
the cloud forest orchids have evolved methods of reproduc-
tion that may seem bizarre but are actually no more unusual
than means which seem more familiar to most humans.
Some of these orchids, for instance, physically resemble the
females of various insects, such as bees and flies; the male

The male quetzal (Pharomachrus mocino), right, *distinguished from the female*, above, *by long plumes and brighter colors, whistles loudly when other quetzals trespass on its feeding grounds. The male sings beautifully during spectacular courting flights.*

175. *Colorful butterflies abound in the mountains of tropical America. Many of them, such as the owl butterfly (Brassolidae), have spectacular protective markings, such as spots on the wings which resemble eyes of predatory animals and frighten off possible attackers.*

176-177. *The skipper (Astrapetes alardus), seen here on the orchid* Catasetum macrocarpum, *is one of several striking butterflies that inhabit the low mountains of Trinidad, just off the coast of Venezuela.*

insects attempt to mate with the flowers and thus spread
the pollen, fertilizing new plants and splashing the gorgeous
colors of the orchid family throughout the misty
highlands.

Orchids are not the only source of blazing colors in the cloud
forests, for these forests shelter an abundance of bright-
hued birds. In the northern Andes, the male scarlet cock-of-
the-rock (*Rupicola peruviana*) performs spectacular dances
on the forest floor to attract a mate. All along the cordillera,
hummingbirds flit from flower to flower like living jewels.
And for more than 1,500 kilometers along the peaks of
Central America, the quetzal (*Pharomachrus mocino*) adds
its gorgeous colors to the misty woods.

The national bird of Guatemala, and a deity to the Aztecs,
Toltecs, and Mayans, the quetzal is the subject of myth and
lore with origins lost in time. The mysterious god of pre-
Columbian Middle America, Quetzalcoatl, was pictured as
either a man wearing a crown of quetzal plumes, or a ser-
pent adorned with them—the legendary "feathered ser-
pent." The name quetzal stems from the Aztec word *quet-
zalli,* used to describe the tail plumes of the male, which are
more than a meter long, even though the bird that carries
them is no larger than a rock dove. Plumes are not the only
lavish adornment of the male quetzal, for the bird has a
vivid red breast and bright green back and head. Atop the
head, the quetzal wears a yellow-green crest.

An eater of fruit, the quetzal nests in holes in trees that
belong to a member of the mulberry (*Morus*) family known
as the masico. Its nest and young suffer considerably from
the widespread burning of the forest to make way for fields,
a practice of the primitive agriculturalists of the quetzal's
mountain home. Quetzals are also killed for their plumes,
even though the trade is prohibited. Unless large tracts of
cloud forest are set aside for the quetzal and protected, this
treasure of the bird world may vanish from the cloud
forests.

The Monarch's Secret Haven

More than 820 meters up the slopes of the Sierra Madre
north of Mexico City, the forest blossoms every November
with orange and black, as millions of monarch butterflies
(*Danaus plexippus*) from the north arrive after one of
the most astounding migrations in the animal kingdom.
The monarchs begin their journey as summer ends, starting
as far north as southeastern Canada. Fluttering on delicate wings that somehow carry them for thousands of
kilometers, the butterflies—traveling as individuals—skim
over the sea waves just off the coast of New England, pass
above the cornfields and prairies of the American midwest,
and the plains of Texas until they reach their Mexican
destination. Discovered there only in 1975, the wintering
haven of the monarch butterflies from eastern North
America was a mystery that had long fascinated scientists.
Once located, the winter sanctuary seemed almost more
astonishing, for the butterflies literally blanket the forest
there.

Not all of the monarchs migrate to the Mexican mountains.
Scientists have found that those which are found there
seem to have hatched late in the summer, and are not sex-
ually mature in time to mate until the following year.
Therefore, they head south to escape the killing cold, and
return in the spring to carry on their kind.

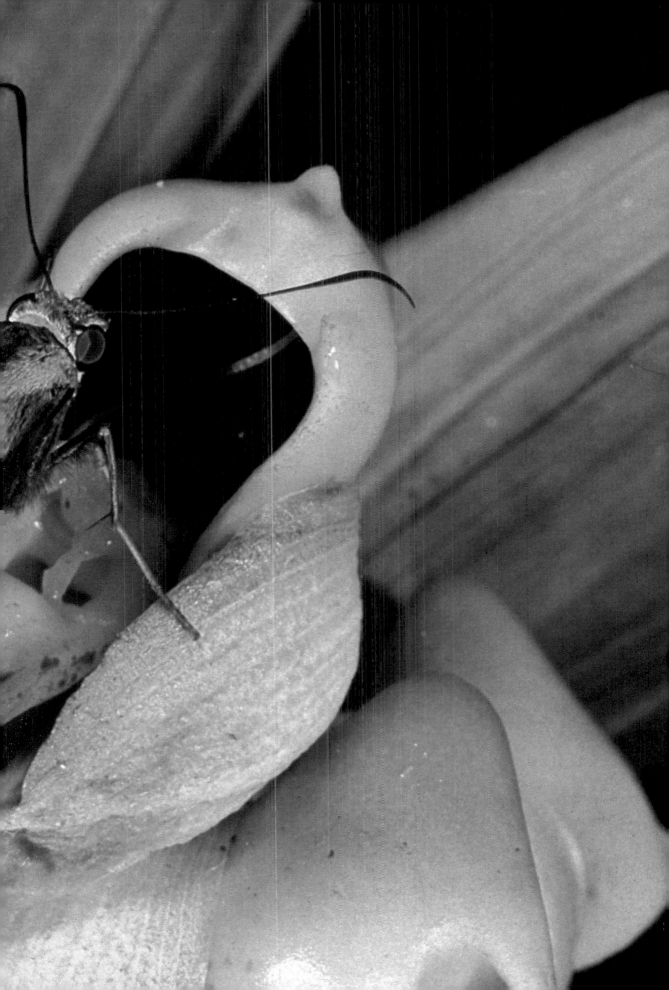

Bear of the Cloud Forests

The cloud forests of the Andes, up to 3,000 meters in altitude, are the home of the mysterious spectacled bear (*Tremarctos ornatus*), the only species of bear in South America. Approximately 140 kilograms in weight, sometimes considerably smaller, this black bear gets its name from the white circles around its eyes. It is largely herbivorous, feeding on fruits, roots, and leaves, and it dearly loves the unfurled young leaves of palm, which it obtains by ripping open palm stalks. On occasion, the bear departs from its vegetarian habits and roams the open areas above the forest to hunt prey such as guanacos (*Lama guanicoe*) and vicuña (*Vicugna vicugna*). The spectacled bear is most at home within the forest, however, and is an agile climber. In fact, it builds a sleeping nest high in the trees, a habit shared among bears only by the small Malayan sun bear (*Helarctos malayanus*) of the jungled hills of Southeast Asia.

There are many unique animals and plants in the cloud forests of Latin America. Seventy-five percent of the plants growing in the Guiana Highlands of northeastern South America, for example, grow only there, another dramatic demonstration of how isolation causes diversity in plants and animals.

Substantial cloud forests grow on the hills of the Paria Peninsula of Venezuela, and atop the low—not more than 1,100 meters high—mountains of northeastern Brazil. The cloud forests of the Andes grow mostly in the northern parts of the 8,000-kilometer-long chain, which stretches from the Caribbean coast of Venezuela to Tierra del Fuego, where as bare rock it recedes into the sea.

In the High Andes

North of Peru, the Andes are exceptionally humid, and thick, moist forests ascend both flanks. Here, between altitudes of 2,000 and 4,000 meters, lives the slim mountain tapir (*Tapirus pinchaque*). Rare, and much less known than its relatives of the lowland jungles, the mountain tapir has a wooly coat, which it surely needs, for it occasionally makes forays as high as the snow line.

Above the forests of the northern Andes tower many snow-capped volcanoes, some of which are active, such as Cotopaxi, which looms above the Ecuadorian capital of Quito. In Ecuador, the equator reaches its highest altitude, crossing a windswept glacier 4,876 meters up toward the peak of the Cayambe Volcano.

The environment of the central Andes, in Peru, Bolivia, and northern Chile, is observably quite different from that of the northern portion of the chain. The western flank of the central Andes is largely dry—bone-dry, in fact—while the eastern slopes are heavily forested with jungle, or "selva," up to 1,000 meters, montane forests from there to 2,000 meters, and cloud forests from 2,000 to 3,500 meters. The difference between the two sides of this section of the mountain chain is due to the direction of the prevailing winds, which blow westward across the midsection of the continent, and as they sweep up the eastern slopes of the mountains are robbed of their moisture. The eastern side—or "Ceja de la Montana"—gets bountiful rainfall and is therefore rich in plant species, while the Pacific slope is so dry that in some parts of it, rain has fallen only once or twice in human memory.

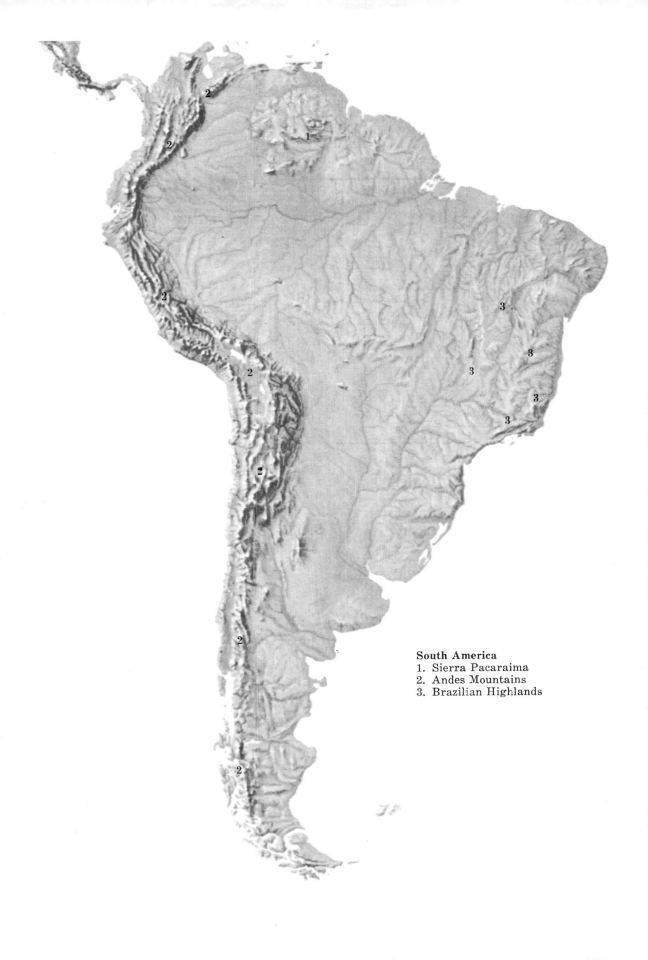

South America
1. Sierra Pacaraima
2. Andes Mountains
3. Brazilian Highlands

180-181. *Llamas* (Lama glama), *domesticated long ago by the Indians, have been prime beasts of burden in the Andes. Like other members of the camel family, they are rather independent in nature and at times cantankerous. When angry they spit their stomach contents at their adversary.*
Below. *Like all other members of the camel family, the vicuna* (Vicugna vicugna) *has a stomach with three chambers and it chews its cud. Less than a meter high and 50 kilograms in weight, the vicuna is the smallest of the family. Young vicunas are born in February after an eleven-month gestation period. The youngsters often graze lying down.*

Because the Andes are a young range—100 million years old —they are saw-toothed and rugged, interlaced with steep valleys, and topped by sheer precipices. This makes for what scientists have called a "patchwork of habitats" close together but isolated and vastly different from one another. The central Andes, where the range reaches its widest point, 700 kilometers from east to west, display a great diversity of conditions at lower altitudes, but higher up are distinguished by a landscape peculiar to the region, a high, cold plateau, approaching tundra in character, sometimes called the puna, or altiplano.

Dry, because of a short rainy season, and scoured by wind, the plateaus above 3,500 meters in the central Andes are at the same time the part of the range most densely populated by people. Four million Peruvians—33 percent of the population of the country—live in the puna, as do 3.5 million Bolivians—more than half of that nation's population. On a bleak landscape dotted with bunchgrass and sedges, farmers grow potatoes and other crops, although the high grasslands are much more suited to grazing.

No wild goats or sheep feed on the high grasslands, or anywhere in South America, for that matter. Neither group reached the southern continent during its invasion by animals from the the north after the gap between North and South America was bridged by land in the Pliocene era. The camel family, however, made the crossing, and two of its members have taken the place of wild goats and sheep in the Andean highlands. The high plateaus are the home of the slim, graceful vicuña and of the guanaco, ancestor of the domestic llama (*Lama glama*) and alpaca (*Lama paca*). Once, great herds of both wild species were common, but they have been woefully reduced, although protection in recent years has kept the vicuña and guanaco from vanishing altogether.

A Varied Fauna

Scattered here and there on the bleak Andean plateaus are large lakes, some of them saline. Typical of these sky-high bodies of water is Laguna Colorada, which stands at an altitude of 4,414 meters in Bolivia, just over the border from Chile. In the shallow waters of the lake feed two rare birds, which were never seen in the outside world until 1960, when a New York Zoological Society expedition collected some and brought them back to the United States. These birds of the high Andes are flamingoes, the Andean (*Phoenicoparrus andinus*) and James's (*P. jamesi*) species, which are found only in the highlands of southern Peru, Chile, and Bolivia. Their presence adds to the rich variety of animal life that is distributed down the backbone of South America.

Flying over the mountains and down the slopes to sea level is the huge Andean condor (*Vultur gryphus*), which weighs 13.5 kilograms and is not only larger than its California counterpart but one of the largest of all flying birds. Only some of the albatrosses (*Diomedea*) have a greater wingspan than the 3-meter spread of the Andean condor, and its wings are much broader than those of the sea birds. The condor is so bulky it launches itself into the air only with great difficulty, and to become airborne needs a runway of several meters. Once aloft, however, it is a picture of aerial grace, soaring with wings hardly moving and its keen eyes sweeping the countryside for carrion.

Above; left. *The Andean condor
(Vultur gryphus) can soar
overhead for hours while it
searches for food. It has excellent
eyesight and is able to spot a
carcass far below. Condors also
have a very good sense of smell.
The male condor is distinguished
by a large wattle.*

184-185. *On the roof of South
America, in shallow brackish
lakes, great flocks of flamingoes
(Phoenicopteridae) breed and
feed. Flamingoes range from the
Andes to Tierra del Fuego. They
feed with their upper bill near
the lake bottom, sieving tiny
plants and animals from the
water. Laguna Colorada, a large,
remote lake, is one of the main
flamingo havens in the Andes.*

Living high among the rocky slopes of the Andes, the mountain viscacha (Lagidium pervanum), top, *and the chinchilla* (Chinchilla laniger), bottom, *have a varied diet consisting of mosses, grass, lichen, and virtually any type of scant vegetation available. They live in colonies, often numbering several dozen animals.*

187 top. *Deer such as the* Pudu pudu *range the Andes forest from Colombia to the southernmost part of Chile.*

187 bottom. *In southern South America the rain-bearing winds blow from the Pacific. The southern Andes block the winds, creating the desert-like region of Patagonia which lies below the slopes of these Argentine Andes.*

Unlike the California condor, however, the Andean species does not restrict its diet to carrion. It occasionally preys upon live animals, such as fawns and, especially, on the sea birds that inhabit the rocky fringes of the Pacific coast of South America. From nests in the mountain heights the condors fly to the sea bird colonies and raid them. Although its talons are weak, the condor can kill its prey with its huge, hooked bill.

Ranging from Tierra del Fuego to Ecuador, the condor is a sociable bird among its own kind, and often gathers in flocks, except in the breeding season. The male condor is distinguished by a large comb and wattles, the female by eyes that become red in the adult, in contrast to the brownish eyes of her mate. The condor pair build little or nothing in the way of a nest, but instead place their one or two eggs on rock, preferably covered with sand.

An animal characteristic of the higher elevation in the Andes below the snow line is the guemal, or Andean deer (*Hippocamelus antisiensis*), which sometimes climbs above 5,000 meters but generally stays at somewhat lower elevations, especially in winter. During the summer the guemals remain around the timberline, in the low thickets of stunted southern beech (*Nothofagus*), a tree which further south and at lower altitudes forms large forests. Guemals are nocturnal in habit, hiding during the day and emerging after dark to feed on leaves, lichens, and mosses. These deer have become rare in many places, due to overhunting and a susceptibility to certain diseases contracted from domestic cattle. The introduction in the Andes of European red deer (*Cervus elaphus*), fallow deer (*Dama dama*), and axis deer (*Axis axis*) from India have added to the troubles of the guemal, for these newcomers have pushed the native deer from much of its habitat.

Several interesting rodents inhabit the Andean plateaus and even the steep slopes above them. The mountain viscacha (*Lagidium pervanum*) and chinchilla (*Chinchilla laniger*) go highest of all, the latter reaching 6,000 meters. Few mammals, in fact, dwell at higher altitudes than these two rodents, which are so prized for their furs. Chinchillas and mountain viscachas live in colonies, among cracks, crannies, and crevices in rocks. Colonies may have scores of members, but both species have been hunted so excessively that big groups are not common. The chinchilla displays particularly unusual behavior for a rodent: it mates for life, which in the wild can be over ten years.

Among the other rodents of the high Andes are the chozchoris (*Octodon degus*), a rat-like creature with a long, rather bushy tail, dark back, and snow-white fur on its underparts; the rock rat (*Aconaemys fuscus*), which creates labyrinthine systems of burrows just beneath the surface of the ground; and the chinchilla rat (*Abrocoma bennetti*), whose skin the people of its region sometimes market to tourists as that of the chinchilla. Inhabiting the plateaus from the central Andes all the way south to the subarctic landscape of Tierra del Fuego is the tucu-tuco (*Ctenomys*), a small, stout rodent that resembles the North American pocket gopher (Geomyidae). Large-headed, with small eyes and ears, the tuco-tuco lives much like the gopher, in long, shallow burrow systems, which the tuco-tuco seems loath to leave. Because its eyes are almost level with the top of its head, the tuco-tuco can survey the countryside around its burrow without projecting its body above the ground.

To the Subarctic South

In the southern part of the tuco-tuco's range, the Andes begin to change. Gone are the high plateaus. Instead, the landscape is grooved by many valleys, and the mountain summits are increasingly lower in altitude as the Andes march toward the sea at Tierra del Fuego. At the same time, however, the snow line moves down the mountainside, until at the tip of the continent glaciers begin at an altitude of only 700 meters. To the east of the mountains stretches the bleak, dry plateau of Patagonia, roved by the rhea (*Pterocnemia pennata*), the giant bird sometimes called, improperly, the South American "ostrich." After the ice ages, the Patagonian plateau was wet and humid, and covered not by sparse grasses but forests of southern beech, like those on the lower slopes of the mountains today. The advance of the forests, which began approximately 9,000 years ago, peaked after some 7,000 years, almost at the edge of the Atlantic Ocean. Then the climate began to change toward the dry conditions of today, and the forest began to retreat before the grasslands, which still goes on today. At Tierra del Fuego, only a few old groves of beech trees are extant, pushed up the low but rugged mountains, and surrounded by grass, in a realm of storms where the Andes disappear into the sea.

From the Mountains to the Sea

The meeting of the Andes and the sea has a symbolic quality, because the mountains of the world are inextricably linked to the sea by an endless round of interactions that have profound influence on the earth and its life. The sand returned to beaches by the ocean waves is composed of rock worn from the mountains by water, while sediment from the sea is the stuff of which many mountains are built. From the mountain heights, great rivers such as the Amazon, Hudson, and Rhone begin their journey to the sea, carrying with them the nutrients from the land that nourish the life of the waters. Sea and mountains are joined by the water cycle—water flows down the slopes from its source in the glacial heights, reaches the sea, evaporates from its surface, and later falls as rain, snow, fog, or mist, in never-ending testimony to the unity that binds all of nature.

188; below. *Darwin's rhea* (Pterocnemia pennata) *inhabits both the flatlands of Patagonia and the heights of the Andes. The male incubates the eggs and acts as guardian of the young.* 190-191. *Improper agricultural practices, such as overgrazing, are rapidly destroying the environment of the altiplano and other parts of the Andes. Because conditions there are so harsh, the altiplano environment is especially vulnerable to destruction.*

APPENDICES

Glossary

Location of geological forms shown in photographs are given in italics folowing entries.

Alluvial fan. A large, open fan-shaped river deposit created at the foot of a mountain by a stream that loses its velocity as it flows off the mountain. *Alluvial fan, European Alps.*

Alpine glacier. A glacier that forms in and flows down a mountain valley. Alpine glaciers usually develop in cirques, and are sometimes referred to as mountain or valley glaciers. *Alpine glacier, Jostedalsbreen Glacier, Norway.*

Anticline. Upward fold in rock resulting from forces within the earth's crust.

Arête. A steep-walled mountain ridge or divide caused by glaciation. *Arête near Reykjavik, Iceland.*

Avalanche. The rapid movement or tumbling of large masses of snow, ice, rock, or soil (mud) down mountain slopes. Snow avalanches occur regularly in high mountain regions, sometimes with velocities of more than 300 km/hour. Although ice avalanches primarily occur on glaciers, they also occur along mountain slopes. Rains often initiate and supply the mobility for soil (mud) avalanches. Earthquakes are the trigger mechanism of most destructive avalanches. In 1963 a rock avalanche caused the deaths of more than 2,500 people in the Italian Alps.

Basalt. Dark-colored, fine-grained volcanic (igneous) rock.

Alluvial fan

Alpine glacier

Arête

Bedrock

Bedrock

Bedrock. All solid rocks at or near the surface of the earth. *Bedrock, Bryce Canyon, Utah.*

Bergschrund. A large crevasse situated between the upper end of a mountain glacier and bedrock above it. Bergschrunds are usually snow-filled in winter but are open (empty) in summer.

Braided stream. A stream that contains an interweaving of small, numerous, intricate channels separated from each other by sand bars and islands. Commonly found at the edge of melting glaciers.

Caldera. A circular depression (larger than a crater) at the summit of a volcano formed by explosion and subsidence or collapse of the volcanic walls.

Canyon. A valley with long, narrow, steep walls and cliffs. A stream or river usually flows along its floor. Larger than a gorge. The most famous canyon is the Grand Canyon in western North America. *Canyon, Big Bend National Park, Texas.*

Cinder cone. A cone-shaped volcano composed of volcanic dust (cinders) and ash.

Cirque. A semicircular or horseshoe-shaped hollow formed along the flanks of glaciated mountain tops and slopes. The depression is produced by glacial plucking, abrasion, and movement. During the winter, cirques may be filled with snow or ice. A small lake (tarn) may form in the cirque during the summer months. *Cirque, Glacier National Park, Montana.*

Canyon

Cirque

Columnar jointing. A pattern of hexagonal, pentagonal, or parallel columns formed from the cooling of lava. *Columnar jointing, Japan.*

Composite volcano. A volcano built and composed of alternating beds of lava that are spilled out in relatively quiet eruptions, and of broken rock materials (pyroclastic debris) deposited through more violent volcanic eruptions. This type of volcano has slopes ranging from 5° at the base to 30° at its summit.

Conglomerate. A sedimentary rock composed primarily of rounded particles the size of gravel, cobbles, and pebbles. *Conglomerate, Italy.*

Continental drift. A general term proposed by the South African geologist Alfred Wegener in 1912 to describe his theory that continents—once part of a single land mass—can move and have moved and rotated away from each other. Wegener's theory was discredited for many years because it provided no viable mechanism to account for such movement. But in the early 1960s concepts of seaflooring spreading (movement of continents away from each other along mid-oceanic ridges on the seafloor) and plate tectonics (the concept of the earth's crust being composed of a large number of plates which float about on a soft substratum in the upper mantle) furnished a reliable method of accounting for displacement of continents.

Columnar jointing

Conglomerate

Convection current. The very slow movement, due to density differences, of huge masses of rock material within the earth's mantle. Convection or density currents are generated at the mantle/core boundary. At this point the lower portion of the mantle is heated and rock materials that are lighter in density than surrounding rock begin to rise. When a current reaches the mantle/crust boundary it cools, becomes heavier, and sinks back into the mantle replacing rising particles. This mechanism is very similar to the process of boiling a pot of water, i.e., the heated water at the bottom of the pot rises because it's lighter, and is replaced with cooler, heavier water from the surface. Theoretically, convection currents occur in pairs, reaching the surface of the earth along mid-oceanic ridges and returning back into the mantle at oceanic trenches located at the edges of continents. These currents are used to explain the hypotheses of continental drift, plate tectonics, and sea floor spreading.

Cordillera. A system of parallel mountain ranges or chains, e.g., the mountain ranges in the western portion of North America from the Rockies to the Pacific Ocean.

Core. Innermost or central portion of the earth which is surrounded by the mantle.

Cornice. A curving, overhanging mass of snow or ice on a mountain ridge.

Crater. An inverted steep-walled, funnel-shaped depression usually located at a volcano's summit. It is through the crater opening that volcanic gases and lava emerge. Craters are created

Crater

as volcanic materials form
about a vent producing a
wall which grows upward as
the volcanic cone is enlarged
by repeated eruptions.
Craters may often reach a
depth of many hundreds of
meters. *Crater, Vesuvius,
Italy.*

Crevasse.   A vertical
V-shaped fracture or crack
in a glacier, generally
ranging in depth from 10 to
200 meters and varying
from several centimeters to
several meters wide.
Crevasses result from the
differential movement
within a glacier as it moves
over jagged surfaces.
Crevasse, France.

Diastrophism.   A term used
to describe all processes that
deform the surface of the
earth. It includes such
processes as folding and
faulting.

Divide.   A high narrow
bridge of land separating
river systems or drainage
basins that flow in opposite
directions from each other.
A divide is the point over
which no water flows. The
most famous is the Conti-
nental Divide in the western
United States.

Dome mountain.   When
molten rock is extruded
from an opening deep in the
earth but does not penetrate
the surface of the earth, the
crust above the molten rock
blisters but does not break.
A bulge forms, hardens, and
generally erodes, resulting
in structures like huge
domes.

Crevasse

Dyke

Erratic

198

Dyke (Dike). Any igneous rock which intrudes through the parallel structures of a neighboring rock. Dykes range in thickness from a few millimeters to tens of meters thick. *Dyke, England.*

Epeirogeny. The processes by which the earth's crust is deformed.

Erosion. The process by which the surface of the earth is worn down. The eroded materials are transposed, and eventually deposited at a new location. Major agents of erosion are wind, streams and rivers, ice (glaciers), and gravity. *Erosion, Grand Canyon, Arizona.*

Erratic. A rock fragment such as a boulder that has been transported glacially and deposited away from its original source. *Erratic, Yosemite National Park, California.*

Escarpment. A clifflike ridge formed either by the erosion of inclined strata of hard rocks or by faulting. Escarpments usually separate lower, sloping areas from higher areas. *Escarpment, near Colorado National Monument, Colorado.*

Exfoliation. The process by which a rock peels apart, forming onion-like layers on the surface of the rock. This flaking process is produced by release of differential stresses within the rock and by rapid temperature changes. *Exfoliation, Yosemite National Park, California.*

Erosion

Escarpment

Exfoliation

Fault. A fracture in the earth's crust accompanied by displacement of one side of the fracture with respect to the other. Faults can be a few centimeters or hundreds of kilometers in length. If the occurring movement is horizontal and parallel to the plane of the fault, the fault is called a "strike-slip fault;" if movement is vertical along the plane of the fault the result is a "dip-slip fault." *Fault, Israel.*

Fault

Fault-block mountains. Mountains produced by large-scale faulting. Fault-block mountains are rather easy to identify because one side of the mountain is steep while the opposite side slopes gently away to a flat surface. Many geologists simply refer to these mountains as block mountains. Classic examples are found in Switzerland, Mongolia, and western portions of the United States. *Fault-block mountains, Crater Lake, Colorado.*

Fjord. A long, submerged, U-shaped glaciated valley extending out into the sea along a mountainous coast. The most famous examples of fjords are in Norway. They resemble "arms of the sea." *Fjord, Geiranger Fjord, Norway.*

Fault-block mountains

Fold. Bends produced in rock by forces operating after deposition or consolidation of the rock. Folds vary in size from millimeters to hundreds of kilometers. Folds can be found in all rock types but are best displayed in sedimentary rocks. *Fold, Israel.*

Fjord

200

Fold mountains. Mountains produced by the folding and buckling of great masses of sedimentary rock. *Fold mountains, Appalachians.*

Foliation. Minerals within a rock that have been structured in parallel arrangements, giving the rock a banded appearance.

Fumarole. A small hole or vent in the earth's crust through which hot gases (principally steam and carbon dioxide) are thrust up. The gases may reach temperatures as high as 1000°C. Fumaroles are usually found in volcanic areas. A famous group is that in the Valley of Ten Thousand Smokes in Alaska. *Fumarole, Solfatara, near Pozzuoli, Italy.*

Geosyncline. A large trough-shaped depression in which many thousands of meters of sedimentary and volcanic rocks have accumulated, and resulting in large scale downbuckling of the earth's crust.

Glacier. A large mass of ice, formed by the compaction and recrystallization of snow, which moves under its own weight. Our most common glaciers are found along the slopes of tall mountains and in their valleys. However, the largest glaciers are huge sheets of ice-covered regions, such as Antarctica and Greenland. Glacial movement is from one centimeter to one meter or so per day. A glacier's size is determined by the amount of fresh snow at its upper level and the temperature at its lower level. Rate of movement is greater at the glacier's center than at its margins and greater where ice is thicker and when temperature of ice is higher. Glaciers are present on

Fold

Fold mountains

Fumarole

mountains on which more snow falls in cold seasons than melts in warm seasons. Glaciers cover approximately 10 percent of the earth's surface, and are found on all continents except Australia. *Glacier, Tongass National Forest, Alaska.*

Gneiss.   A highly foliated metamorphic rock with alternating bands of light and dark minerals. *Gneiss, Norway.*

Gorge.   A narrow passage, or a narrow deep valley flanked by steep walls and usually created by the wearing action of a stream. *Gorge, Urubamba Valley and Urubamba River, Peru.*

Graben.   A depressed portion of the earth's crust bounded on at least two sides by faults. The Great Rift Valley of Africa and the Rhine Graben of Germany are the result of graben depressions.

Granite.   A coarse-grained igneous rock containing feldspar and quartz. Granite is one of the earth's hardest rocks. *Granite, Austria.*

Horn.   A steep pyramidal-shaped glaciated mountain peak. The Matterhorn in the Swiss Alps is a well-known example.

Horst.   A relatively long stretch of land that has been thrust upward between two faults.

Ice-fall.   The point at which a glacier passes over a particularly steep drop in the bedrock on which the glacier rests. At this point the glacier is often cut extensively by crevasses. *Ice-fall, Briksdalsbreen Glacier outlet, Norway.*

Glacier

Gneiss

Gorge

Igneous rock. Any rock that has solidified from a semimolten state. Igneous rocks are one of three primary families of rocks. The other two are sedimentary and metamorphic. Some of the most common igneous rocks are granite and basalt.

Joint. Any natural crack or break in a rock along which there has been no movement. Joints commonly occur in parallel cracks and appear to give a pattern of foliation to many rocks. *Joint, Middlehead, Nova Scotia.*

Karst. An area created by limestone dissolved by subsurface water and characterized by caves, caverns, and natural bridges. Named after the Karst district of Yugoslavia; also found in many other regions of the world. *Karst, Sedom, Israel.*

Lava. Molten material spilled out of vents and shafts at the surface of the earth; a fine-grained igneous rock forms when this molten material cools and solidifies. Because lava is a poor conductor of heat, it cools slowly; estimates are that a lava flow 9 meters thick reaching the surface at a temperature of 1100°C would cool to air temperature in about 10 years.

Limestone. A sedimentary rock composed chiefly of calcium carbonate.

Magma. Molten or fluid material within the earth from which an igneous rock results by cooling.

Mantle. The part of the earth situated between the earth's crust and the core of the earth.

Granite

Ice-fall

Joint

Karst

Massif. A mountain mass broader than a range, often consisting of plateau-like areas from which prominent peaks rise.

Mesa. A flat-topped, isolated mountain usually composed of sedimentary rock and flanked on at least one side by a steep cliff. *Mesa, Masada, Israel.*

Metamorphic rock. Rocks that have been transformed mineralogically, structurally, or chemically through application of tremendous heat, pressure, or chemical fluids. Gneiss, schist, slate, and marble are metamorphic rocks.

Moraine. Unsorted debris and rock fragments carried by a glacier and deposited into mounds and/or ridges. *Moraine, near Montauk Point, Long Island, New York.*

Moulin. A vertical hole in a glacier carved by a stream that has extended to the bedrock below.

Mountains. An elevated portion of the earth's crust higher than a hill and having a conspicuous summit, steep sides, and outcroppings of bare rock. Mountains are formed by volcanic activity and earth movements and through erosional activity. *Mountains, near Reykjavik, Iceland.*

Nappe. Huge masses of rock, thrust or folded over adjacent areas of the earth's surface. The Swiss Alps are remnants of nappe structures.

Mesa

Moraine

Névé. The area on a glacier which is always covered by snow. The term "névé" is also used by many geologists as an equivalent to the British term "firn." When used in this manner its definition is restricted to explanation of the process that converts snow into ice by means of alternate freezing, melting, sublimation, condensation, and pressure.

Mountains

Nuée Ardente. A hot, sometimes incandescent cloud of gases that erupts from a volcano and flows downhill. In 1902, 30,000 people were killed in the West Indies by a nuée ardente as it roared down Mount Pelée.
Mount Pelée. *Nuée Ardente, Mount Pelée, West Indies.*

Obsidian. Volcanic glass that is generally dark or black with conspicuous concoidal fracture. It is formed by fast-cooling lava.

Orogeny. The process of the formation or making of mountains.

Outcrop. Exposed rock at the earth's surface. *Outcrop, Colorado National Monument.*

Piedmont glacier. When two or more alpine glaciers coalesce at the foot of a mountain, a mass of ice called a piedmont glacier is formed.

Plateau. A large elevated (300 meters) piece of land (larger than a mesa) that has a flat top and is bordered on at least one side by a steep slope.

Nuée Ardente

Outcrop

Plate Tectonics. The theory which suggests that the earth's crust and upper mantle have been broken into large, thin (50–250 km. thick) blocks or plates which are in constant motion. Plate displacement and movement is caused by magma and lava rising from the mantle to the earth's surface at the intersection or boundary of adjacent plates. The energy of the rising igneous materials nudges the plates apart and when the materials solidify, the plates are enlarged, forcing them to shift and readjust position. Many times plates collide and one is forced under an adjacent plate. This process is called subduction and usually occurs in the vicinity of oceanic trenches and island arcs.

Pumice. A lightweight volcanic glass, usually white or gray in color. This rock contains so many holes (vesicles) that it may float in water.

Relief. A topographic term describing the difference in elevation between the highest and lowest points in a region.

Rift. A zone in the earth's crust associated with fractures, faults, volcanic activity, and earthquakes. *Rift, Red Sea, Israel.*

Roches moutonnées. Rocks that are ground, sculpted, and rounded by glacier action into elongated masses oriented in the direction of ice movement. These rocks are easily recognizable, as one side is smoothly sloped and rounded and the other steep and stubbly. *Roches moutonnées, upper New York State.*

Rift

Roches moutonnées

Sedimentary Rock

Sedimentary Rock

Saddle. The lowest point on a ridge between two peaks.

Schist. A metamorphic rock composed of foliated light and dark-colored minerals, mostly mica flakes arranged in parallel patterns.

Sea-floor spreading. A general term explaining the growth of sea floor (1–10 cm. per year) as magma rises along the midoceanic ridges and spreads laterally away from the ridge. The older sea floor is absorbed between adjacent plates in the crust. Sea-floor spreading is the major mechanism for the hypothesis of continental drift and plate tectonics.

Sedimentary rock. Rocks formed through the accumulation and consolidation of loose materials called sediment. Nearly all sedimentary rocks are layered (stratified). Some of the more common sedimentary rocks are sandstone, conglomerate, limestone, and shale. *Sedimentary rock: right, Israel; far right, Norway.*

Shale. A sedimentary rock composed primarily of mud, silt, and clay.

Shield volcano. A volcano composed primarily of basaltic lava flows, with slopes ranging from 2° at the base to 10° at the summit. The major volcanoes on Hawaii are shield volcanoes.

Sill. A tabular igneous rock that has injected itself between two sedimentary layers. *Sill, Banks Island, Northwest Territory, Canada.*

Sill

Snow line. The line above which snow remains year-round. Altitude of snow line on a given mountain is determined by latitude as well as annual rainfall.

Subduction Zone. The part of the earth's crust in which one plate is forced under another plate. The regions of subduction zones are associated with island arcs and oceanic trenches.

Syncline. The downward fold in rock resulting from forces within the earth's crust.

Talus. Loose piles of broken rock, boulders, and gravel derived from and accumulating along the slopes of hills, mountains, and cliffs. *Talus, Yosemite National Park, California.*

Tarn. A small deep mountain lake located in a cirque. *Tarn, Norway.*

Terminal moraine. The moraine that marks the farthest point of a glacier's most advanced position.

Timberline. The point on a mountain above which no trees can grow. In a regional sense, the latitude above which the trees do not grow.

U-shaped valley. A valley with a U-shaped profile. Nearly all U-shaped valleys are formed by glacial erosion. When a glacier moves through a normal V-shaped stream valley it tends to grind down the valley walls, creating a streamlined U-shaped profile. *U-shaped Valley, Rocky Mountain National Park, Colorado.*

Tarn

Talus

Volcanic plug

Volcanic plug. An isolated mountain or hill of resistant rock. An erosional remnant from the interior of an extinct volcano. Plugs may be over 400 meters high, and are found throughout the world. Shiprock in New Mexico is the most famous example in the United States. *Volcanic plug, Shiprock, New Mexico.*

Volcano. A conical land form with a vent at its summit created by the emission of molten liquids (lava, hot gases, rock fragments, and cinders). Evidence indicates that at some time every part of the world has been volcanic. Usually cone- or dome-shaped, volcanoes are temporary features of the earth; erosional factors eventually destroy them. Famous volcanoes include Vesuvius in the Mediterranean, Krakatoa in the East Indies, Kilimanjaro in Africa, and Mauna Loa in Hawaii.

Water table. The uppermost surface of underground water held (saturated) in the rocks. The level of the water table varies with altitude, increasing in mountain regions and dropping in valleys.

Weathering. The actual process of destruction and alteration of rocks. Chemical weathering is the process through which rocks are altered by the dissolution of minerals within the rock. Mechanical weathering involves actual physical breakup of rock through processes such as temperature changes and alternate freezing and thawing of water in the rock. *Weathering, east of Perth, Australia.*

U-shaped valley

Weathering

209

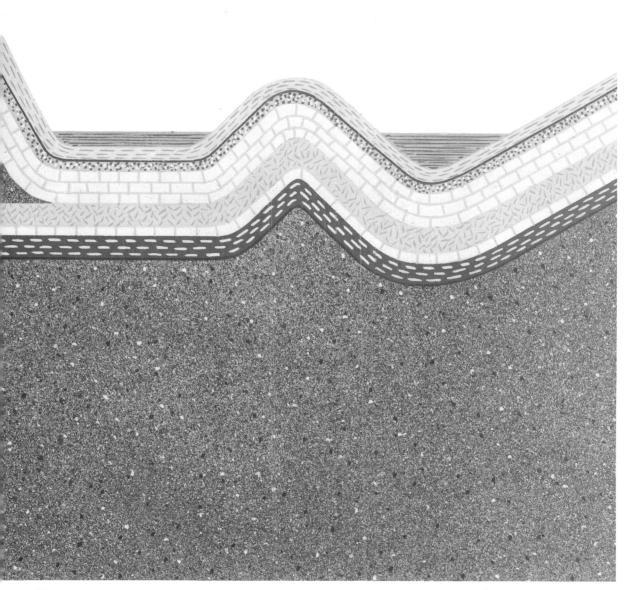

Illustrated here are various mountain building processes.
Left. *Volcanoes are created when lava from deep in the earth is extruded through the crust.*
Center. *Some mountains are formed when liquid rock bulges from the earth but does not break through.*
Right. *Simple folded mountains are caused by a crumpling of the crust.*

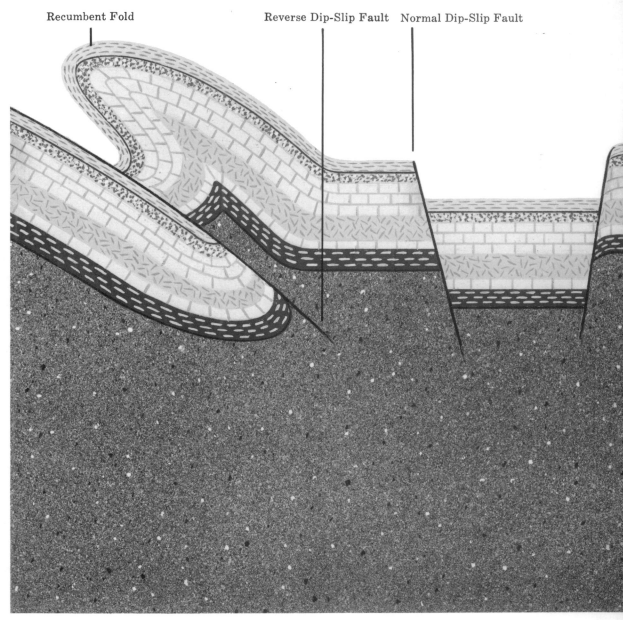

Recumbent Fold   Reverse Dip-Slip Fault   Normal Dip-Slip Fault

Second from left. *When the axial
plane of a fold is inclined to the
horizontal, the fold is known as
"recumbent." Faulting occurs
when a portion of the earth's
crust breaks and an adjacent portion is displaced, creating mountains that are gently sloped on
one side and steep on the other.*
Third from left. *Normal dip-slip
faults occur when the movement
is vertical along the fault's plane.*
Extreme left. *When the motion
is reversed and the angle of the
dip is greater than 45 degrees
the fault is known as a reverse
dip-slip fault.*
Fourth and fifth from left. *Horsts
and grabens are other examples
of dip-slip faults.*
Extreme right. *Strike-slip faults
occur when the movement is horizontal and parallel to the plane
of the fault.*

Horst   Graben   Strike-Slip Fault

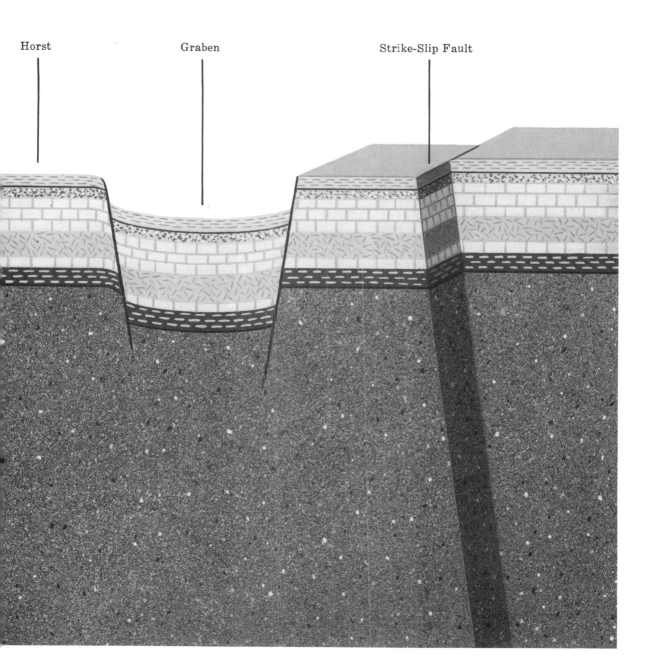

High Places of the World

The high points of all continents and of significant ranges worldwide—including ranges and peaks mentioned in the text—are ranked by range, altitude, and peak.

Asia is the highest, largest, and most diverse of the continents. If Asia Minor is included, Asia contains both the highest (Everest) and the lowest (Dead Sea) points on the surface of the earth. The "roof of the world" is here in the great Himalayas, a mountain range with more than thirty peaks of 7,600 meters or more.

The Caucasus—a range entirely within the Soviet Union—and the Alps contain the highest relief in Europe, a great deal of which is low both in altitude and in relief. The mountains of the central Caucasus, part of the geographical boundary between Europe and Asia, average 1,000 meters higher than their counterparts in the western and central Alps. There are at least fifty mountains in the Alps with summits higher than 4,000 meters.

In proportion to its size, Africa has fewer high mountains and lowland plains than any other continent. Ethiopia comprises the most extensive high area. The East African Plateau is highest (2,400 meters) in Kenya and occasionally contains high volcanic peaks such as those of Ruwenzori or Mountains of the Moon, some of which are more than 5,000 meters high. A well-defined but discontinuous escarpment 3,000 meters high, known as the Drakensberg, runs along the southern part of the east African coast. The only large areas more than 1,000 meters in altitude are the Atlas Mountains of Morocco and the central Saharan massifs of Hoggar and Tibesti. Small areas higher than 1,000 meters are found in Sudan and in volcanic Mt. Cameroon.

The western half of North America, particularly Alaska and the Yukon Territory of Canada, contains the areas of highest relief on that continent and culminates in Mount McKinley. There are nine peaks in the Alaska and St. Elias ranges more than 5,000 meters high and numerous other peaks more than 4,000 meters. The Canadian Rockies to the south contain more peaks higher than 3,300 meters. The Colorado Rockies, still further south, are on the average even higher, with almost sixty named peaks over 4,200 meters and more than one thousand over 3,300 meters.

The eastern half of North America is low and the mountains, ancient and well eroded, known in the aggregate as the Appalachians, parallel the Atlantic Coast. In this system there are less than sixty peaks of 1,200 meters or higher.

Three volcanoes more than 5,000 meters high as well as a few more over 4,000 meters are in Mexico and Guatemala.

South America is dominated by the Andes, which run north/south virtually the entire length of the continent's Pacific Coast and which rank second behind the Himalayas in average height. There are many summits higher than 6,700 meters in the Andes, particularly in the Cordillera Blanca of Peru and along the Chilean–Argentine border west of Santiago, Chile. The other high areas of this continent are to be found on the Atlantic side in the Guiana and Brazilian highlands. These highlands are about 4,000 meters high with some low areas (hills of 2,000 meters) rising, on the southern edge, to more than 2,700 meters in the Guiana Highlands and to 3,000 meters on the summits of the Brazilian Highlands.

Australia, the smallest continent, has no high mountains.

Near the west coast the land rises abruptly into a plateau which covers most of western Australia. In the east the Great Dividing Range culminates in Mt. Kosciusko, the continent's high point. In contrast, the islands of New Zealand are quite hilly.

The altitudes used in this list reflect those of accepted cartographic sources. Altitudes, however, can change and can vary from source to source due to lack of surveys, inaccurate surveys, or over-enthusiastic local authorities. With certain exceptions, such as Mont Blanc, Monte Rosa, and Tien Shan, effort has been made to eliminate the local reference to peak or mountain, i.e., cerro, cordillera, daği, jabel, jebel, koh, kuh, mount, pico, nevado, shan, tau.

All altitudes are given in meters.

Range or Location, Country	Mountain	Altitude
Europe		
Caucasus, USSR	Elbrus	5633
	Dych	5203
	Shakhara (Shakara)	5201
	Koshtan	5104
	Dzangi	5049
Mont Blanc, France/Italy	Mont Blanc	4807
Pennine Alps, Switzerland/ Italy	Monte Rosa	4634
	Dom	4545
	Liskamm (Lyskamm)	4527
	Weisshorn	4505
	Matterhorn	4478
Bernese Alps, Switzerland	Finsteraarhorn	4274
	Jungfrau	4158
	Eiger	3970
Dauphiné Alps, France	Les Écrins	4101
	Meije	3983
Graian Alps, Italy	Gran Paradiso	4061
Bernina Alps, Italy	Bernina	4049
Ötztal Alps, Italy	Ortles	3899
Ötztal Alps, Austria	Wildspitze	3744
Cottian Alps, France/Italy	Viso	3841
Hohe Tauern, Austria	Grossglockner	3798
Sierra Nevada, Spain	Mulhacén	3478
	Valeta	3392
Pyrenees, Spain	Aneto	3404
	Poseta	3371
	Perdido	3352
Dolomites, Italy	Marmolada	3342
Sicily, Italy	Etna	3340
Maritime Alps, France/Italy	Argentera	3297
Bavarian Alps, Federal Republic of Germany	Zugspitze	2963
Rhodope, Bulgaria	Musala (Stalin)	2925
Gran Sasso d'Italia (Apennines), Italy	Corno	2914
Mytikas, Greece	Olympus (Olimbos)	2911
Julian Alps, Yugoslavia	Triglav	2863
Corsica, Italy	Cinto	2707
High Tatras (Carpathians), Czechoslovakia	Gerlach	2663
Cantabria, Spain	Cerrado	2642
Pindus, Greece	Smolikas	2637
Jötunheim, Norway	Glittertind	2470
	Galdhöppin	2469
Urals, USSR	Narodnaya	1894
Dinaric Alps, Yugoslavia	Trebeviç	1629
Grampians, Scotland	Ben Nevis	1343
Harz, Democratic Republic of Germany	Brocken	1142
Africa		
Kilimanjaro National Park, Tanzania	Kilimanjaro	5890
Mt. Kenya National Park, Kenya	Kenya	5199
Ruwenzori, Uganda/Zaïre	Stanley	5110

Range or Location, Country	Mountain	Altitude
Ruwenzori, Uganda	Speke	4890
	Baker	4840
Ruwenzori, Zaïre	Emin	4800
Ruwenzori, Uganda	Gessi	4720
	Luigi di Savoia	4630
Semien (Semyen), Ethiopia	Ras Dashan (Dejen)	4620
	Bwahit	4509
Arusha, Tanzania	Meru	4565
Virunga, Uganda/Rwanda/ Congo	Karisimbi	4507
	Mikeno	4437
	Muhavura	4127
Trans-nzoi/Sebei, Kenya/ Uganda	Elgon	4321
Arusi, Ethiopia	Encuolo	4311
Mendebo, Ethiopia	Batu	4307
High Atlas, Morocco	Toukbal	4165
	Timezgadioune	4089
	M'Goun (Ighil M'Goun)	4071
	Afella	4043
Cameroon, Cameroon	Cameroon	4069
Aberdare, Kenya	Ol Doinyo Lastima	3999
Drakensberg, Union of South Africa	Thabantshonyana	3482
	Inajusti	3408
	Champagne Castle	3371
	Giant's Castle	3316
	aux Sources	3299
Tibesti, Chad	Emi Koussi (Emi Cussi)	3415
	Tarso Ahon	3325
	Toussidé	3265
	Keguer Terbi	3150
	Ehi Tini	3040
Imatong, Sudan	Kinyeti	3196
Réunion, Réunion	Piton des Neiges	3070
Mlanje, Malawi	Spitwa (Mlanje)	3000
Fernando Po Island, Guinea Bissau	Santa Isabel	3000
Hoggar (Ăhaggar), Algeria	Tăhat	2918
	Ilâmen (Ilamene)	2760
	Amdjer	2750
	Ăsēkrem (Assekrem)	2728
Tsarantanana, Malagasy Republic	Maromokotro	2880
Kaokoveld, Namibia	Brandberg	2606
Inyanga, Rhodesia	Inyangani	2596
Serra Upanda, Angola	Moco	2460
Asia Minor		
Elburz (Alborz), Iran	Demavend	5954
Eastern Turkey, Turkey	Ararat (Büyük Ağri)	5165
	Suphan	4434
Zagros, Iran	Zard	4548
Kurdistan, Turkey	Resko Tepe (Cilo)	4168
Lebanon, Lebanon	Qurnat as Sawda	3088
	Hermon	2814
Sinai, Israel	Katrīnah	2637
	Sinai (Musa)	2256

Asia and the South Pacific including Australia and New Zealand

Range or Location, Country	Mountain	Altitude
Nepal Himal, Nepal/China	Everest (Chomolungma, Sagarmatha)	8848
Nepal Himal, Nepal/Sikkim	Kan(g)chenjunga	ca. 8496
Nepal Himal, Nepal	Makalu	8481
	Lhotse	8426
	Dhaulagiri	8167
	Manaslu (Kutang I)	8165
	Annapurna I	8051
Nepal Himal, Nepal/China	Gosainthan (Shisha Pangma)	8014
Karakoram, Pakistan	K2 (Godwin Austin)	8760
	Gasherbrum I (Hidden Peak)	8068
	Broad Peak (Falchan Ri)	ca. 8000
	Khiangyang Kish	7852
	Dhisteghil Sar	ca. 7700
Punjab Himal, India	Nanga Parbat	7816
Pamirs, China/USSR	Kungur	7719
	Kungur Tjube	7595
Pamirs, China	Mustagh Ata	7546
Pamirs, USSR	Kommunisma	7482
	Lenin	7134
Hindu Kush, Afghanistan/ Pakistan	Tirich Mir	7699
Hindu Kush, Afghanistan	Noshaq	7492
Hindu Kush, Pakistan	Istor-o-nal	7389
Arka (Kun Lun), China	Ulugh Muztagh I	ca. 7546
Tien Shan, USSR	Pobeda (Pobedy)	7439
	Tengri Khan	6995
Snow (Irian Jaya, New Guinea), Indonesia	Jaya (Sukarno, Carstensz)	5029
Kamchatka, USSR	Klyucheskaya	4750
Katun (Altai), USSR	Belukha	4506
Hawaii, USA	Mauna Kea	4205
	Mauna Loa	4170
Sabah, Malaysia	Kinabalu	4101
Taiwan, Republic of China on Taiwan	Yu (Morrison)	3997
	Tz'u'kaw (Tzekao, Sylvia)	3884
Naga, Burma/India	Saramati	3826
West Sumatra, Indonesia	Kerenci	ca. 3800
Northern Alps (Honshu), Japan	Fujisan	3776
	Okuhotaka	3190
	Yarigatake	3178
Southern Alps, New Zealand	Cook	3764
	Tasman	3498
	Malte Brun	3176
Southern Alps, Japan	Kitadake	3192
Lao Kai, Socialist Republic of Vietnam	Fan Si Pan	3142
Central Alps, Japan	Kisokomagatake	2956
Mindanao, Philippines	Apo	2954
North Island, New Zealand	Ruapehu	2797
Paetku (Changpai), People's Republic of China/Democratic People's Republic of Korea	Paektu (Changpai)	2744

Range or Location, Country	Mountain	Altitude
Chiang Mai, Thailand	Doi Inthanon	2595
Hokkaido, Japan	Asahidake	2290
Snowy, Australia	Kosciusko	2230
Shikoku, Japan	Ishizuchi	1981
Kyushu, Japan	Kuju	1788
Tasmania, Australia	Ossa	1617
Macdonnell, Australia	Zeil	1510
Sikhote, USSR	Zapovednik	1448
Hammersley, Australia	Bruce	1227
Barlee, Australia	Augustus	1106
Northwest Territory, Australia	Ayres Rock	867

North America (Alaska, Canada and Continental United States)

Alaska, USA	McKinley	6190
	Foraker	5300
	Hunter	4440
St. Elias, USA/Canada	Logan	5951
	St. Elias	5490
St. Elias, Canada	Lucania	5226
	King	5172
	Steele	5068
Wrangell, USA	Blackburn	4996
	Sanford	4949
	Wrangell	4317
Fairweather, USA	Fairweather	4663
Sierra Nevada, USA	Whitney	4418
Colorado Rockies, USA	Elbert	4399
	Massive	4395
	Harvard	4393
Cascades, USA	Rainier	4392
	Shasta	4316
	Hood	3424
Wind Rivers, USA	Gannett	4202
Tetons, USA	Grand Teton	4196
Uintas, USA	Kings	4123
Coast Range, Canada	Waddington	4016
	Tiedemann	3828
Chugach, USA	Marcus Baker	4013
Yosemite, USA	Lyell	3997
Canadian Rockies, Canada	Robson	3954
	Columbia	3747
	North Twin	3684
San Francisco, USA	Humphrey	3851
Selkirks, Canada	Sir Sandford	3533
Aleutian, USA	Torbert	3497
Purcells, Canada	Farnham	3457
Cariboos, Canada	Sir Wilfred Laurier	3429
Monashees, Canada	Monashee	3246
Glacier National Park, USA	Cleveland	3185
Brooks, USA	Isto	2761
Guadalupe, USA	Guadalupe	2667
Olympics, USA	Olympus	2424
Smokies (Appalachians), USA	Mitchell	2037
White (Appalachians), USA	Washington	1917
Adirondacks (Appalachians), USA	Marcy	1629

Range or Location, Country	Mountain	Altitude
Baxter State Park (Appalachians), USA	Katahdin	1605
Alleghenies (Appalachians), USA	Davis	979
New Mexico, USA	Shiprock	488

Mexico, Central America and the Caribbean

Range or Location, Country	Mountain	Altitude
Sierra Madre Oriental, Mexico	Orizaba (Citlaltépetl)	5699
	Cofre de Perote (Nauhcampetépetl)	4537
Puebla/Mexico, Mexico	El Popocatépetl	5452
	Ixtaccíhuatl	5286
Pueblo/Tlaxcala, Mexico	Malinche	4461
Jalisco, Mexico	Colima	4265
Sierra Madre, Guatemala	Tajmulco	4210
	Tacaná	4093
	Acatenango	3976
Talamanca, Costa Rica	Chirripó Grande	3920
Cordillera Central, Puerto Rico	Punta	1338
Luquillo, Puerto Rico	El Toro	1074

South America

Range or Location, Country	Mountain	Altitude
Mendoza, Argentina	Aconcagua	6959
Puna or Northern Andes, Chile	Ojos del Salado	6885
	Llullaillaco	6723
	Toro	6386
	Parinacota (Payachata.Sur or Muru Payachata)	6330
	Tórtolas	6323
Catamarca/La Rioja, Argentina	Pissis	6780
Catamarca, Argentina	Gonzales (sin Nombre)	6660
	Nacimiento	6490
San Juan, Argentina	Mercedario	6770
	Ramada	6410
Blanca, Peru	Huascarán	6769
	Chopilcalqui	6400
	Huantsán	6395
	Huandoy	6395
Salta, Argentina	Liberatador	6720
Mendoza/Santiago, Argentina/Chile	Tupungato	6670
Huayhuash, Peru	Yerupajá	6632
	Siulá	6352
Incahausi, Argentina/Chile	Tres Cruces	6630
	Incahuasi	6620
Occidental, Peru	Coropuna	6615
	Ampato (Hualca-Hualca)	6360
	Solimana	6318
Occidental, Bolivia	Sajama	6520
Real, Bolivia	Illimani	6462
	Ancohuma	6430
	Illampu (Sorata)	6362
Vilcanota, Peru	Ausangate	6384

Range or Location, Country	Mountain	Altitude
Occidental, Ecuador	Chimborazo	6267
	Sangay	5323
	Illiniza	5266
Oriental, Ecuador	Cotopaxi	5897
	Cayambe	5789
	Antisana	5706
Santa Marta, Colombia	Colón	5775
	Bolívar	5775
Cocuy, Colombia	Alto Ritacuba (Ritacuba Blanco)	5464
Huila, Colombia	Huila	5439
Central, Colombia	Ruiz	5239
Merida, Venezuela	Bolívar	5002
	Bonpland	4885
Patagonia, Chile	San Valentín	3876
Patagonia, Argentina/Chile	FitzRoy (Chaltén)	3375
Guiana Highlands, Brazil	Neblina	ca. 3000
Tierra del Fuego, Chile	Yagan	2469

Antarctica

Sentinel, Ellsworthland	Vinson Massif	5139
	Tyree	4965
	Shinn	4801
	Gardner	4688

Rocky Mountain Goat

Bighorn Sheep

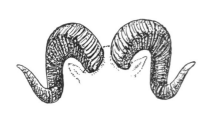

Dall's Sheep

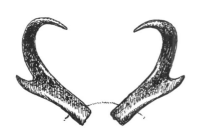

Pronghorn Antelope

Horns

Horns have evolved on ungulates during the past 25 million years. They vary greatly in size and shape from species to species and even among subspecies. The structure of all horns, however, is similar: a core of bone arising from the frontal plate of the skull covered with keratin, the main ingredient of nails, hooves, and hair. While horns serve as weapons against predators, their primary function is as a weapon in ritualized combat between males of the same species for the right to mate. Length of horns given below are approximate for male animals.

Europe
Mouflon (*Ovis ammon musimom*) actual length of horns—75 cm.
Alpine Ibex (*Capra ibex ibex*) actual length of horns—90 cm.
Chamois (*Rupicapra rupicapra*) actual length of horns—25 cm.
Wisent (*Bison bonasus*) actual length of horns—50 cm.
Spanish Ibex (*Capra pyrenaica*) actual length of horns—1 m.

Africa
Abyssinian Ibex (*Capra ibex walie*) actual length of horns—70 to 110 cm.
Nyala (*Tragelaphus buxtoni*) actual length of horns—110 cm.
Bongo (*Taurotragus eurycerus*) actual length of horns—90 cm.
Eland (*Taurotragus oryx*) actual length of horns—113 cm.
Buffalo (*Syncerus caffer*) actual length of horns—1 m.

Asia
Marco Polo Sheep (*Ovis ammon poli*) actual length of horns—190 cm.
Armenian Urial (*Ovis ammon arkal*) actual length of horns—90 cm.
Siberian Ibex (*Capra ibex sibirica*) actual length of horns—140 cm.
Markhor (*Capra falconeri*) actual length of horns—160 cm.
Yak (*Bos mutus grunniens*) actual length of horns—95 cm.
Takin (*Budorcas taxicolor*) actual length of horns—63 cm.

North America
Rocky Mountain Goat (*Oreamnos americanus*) actual length of horns—30 cm.
Bighorn Sheep (*Ovis canadensis*) actual length of horns—90 cm.
Dall's Sheep (*Ovis canadensis dalli*) actual length of horns—90 cm.
Pronghorn Antelope (*Antilocapra americana*) actual length of horns—25 cm.

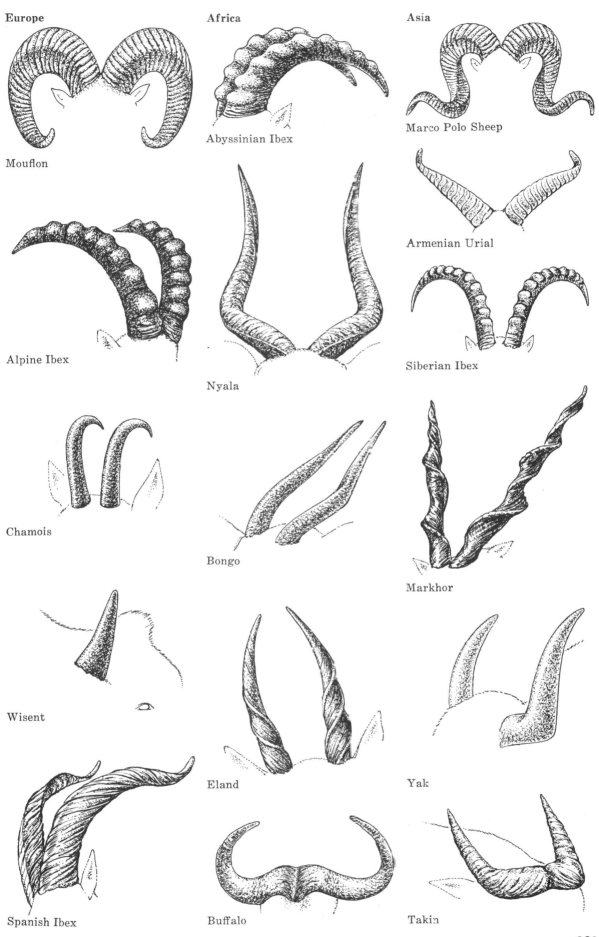

Europe

Mouflon

Alpine Ibex

Chamois

Wisent

Spanish Ibex

Africa

Abyssinian Ibex

Nyala

Bongo

Eland

Buffalo

Asia

Marco Polo Sheep

Armenian Urial

Siberian Ibex

Markhor

Yak

Takin

223

Index

Page numbers in bold face type indicate illustrations.

Aberdare National Park, 115
Aberdare Range, **96**, 105, 106, 108, 114, 115
Abies balsamea, 27, 152
Abies fraseri, 158
Abrocoma bennetti, 186
Abruzzi National Park, 67
Accentor, alpine, 88, 126
Accentor, maroon-backed, 126
Accipiter cooperii, 157
Accipiter striatus, 157
Acer saccharum, 152
Aconaemys fuscus, 186
Aconcagua, **22**
Acrididae, 128
Adirondacks, **148**, 152, 153
Aegolius acadicus, 161
Aegypius monachus, 76, 77
Africa, 96–117
Agkistrodon contortrix mokasen, 161
Ahagger, **96**
Ailuropoda melanoleuca, 121, 128 **132**
Ailurus fulgens, 128, 133
Air plants, 27, 172
Alaska Range, **148**, 165
Alborz Mountains, 121, 138
Alces alces, 48, 168
Alders, 48
Alectoris graeca, **127**
Aleutians, 20
Alleghenies, 21, 157
Alluvial fan, 194, **194**
Alnus, 49
Alpaca, 182
Alpenrose, 28, 31
Alpine glacier, 194, **194**
Alpine zone, 30, 49, 158
Alps, 18, 19, 20, 22, 23, 27, 28, 30, 37, **40**, 41, 47, 49, 53–54, 55, 58, 67, 69, 75, 76, 82, 87, 88, 93, 126, 214
Altai Mountains, **22**, **118**, 121, 135, 138
Altiplano, 182, **190–191**
Altyn Tagh, 121
Amazona vittata, 34
Ambystoma maculatum, **162**
Ambystoma opacum, **162**
Americas, 148–191
Ammotragus lervia, **102–103**
Andes, 18, **18**, 20, 22, 23, **26**, 27, **29**, 174, 178, **179**, 182, 186, 188, 214
Aneides aeneus, **160**
Annamese Cordillera, 118

Anoles, Carolina, 161
Anolis carolinensis, 161
Anser anser, **130–131**
Anser indicus, 128
Antelopes, 11, **104**, 108, **109**, 140
Antelopes, goat, 143
Antelopes, Tibetan, *see* Chiru
Anticline, 194
Apennines, 34, **40**, 49, 53, 55, 66–67, 75
Appalachians, 18, 23, **148**, 149, 152–153, 157, 158, 161, 214
Aquila chrysaetos, 47, **74**, 75, 157
Arabian Mountains, **118**
Arctostaphylos nova-ursi, **50**
Arete, 194
Argali, **29**, 135
Argentine Andes, **187**
Argiopidae, **128**
Arrhenatherum, 158
Arundinaria, 106
Ascaphus truei, 171
Asia, 118–143
Asp viper, **92**
Aspens, 49
Astrapetes alardus, **176–177**
Atelopus, **172**, **173**
Atlas Mountains, 18, 26, **96**, 97, 101, **102**, 105, 214
Australasia, 143
Australia, **120**, 144
Australian Alps, 144
Avalanches, 54, 194
Azaleas, 66, 158

Baboons, **101**
Badgers, 28
"Balds," 158
Balkan Mountains, 20, **22**, 40, 55, 65, 66, 70
Bamboo, 26, **29**, 106, 108, **108**, 128
Barbary sheep, **102–103**
Basalt, 194
Bayan Kara Shan, 121, 128
Bears, 55
Bears, black **29**, 133, **133**, 158
Bears, brown, 34, 47, 53, 67, 68, 69, 133, 168, **168**
Bears, grizzly, 28, 149, 168, **168**, **169**, 170
Bears, Malayan sun, 178
Bears, spectacled, **29**, 180
Bearberry, **50**
Beavers, 158
Bedrock, **194**, 195
Beech, 53, 55, 66, 145, 152, 189

Beech, southern, 149, 186
Beetles, carabid, 114
Beetles, ground, 93
Beetles, leaf, 52
Bergschrund, 195
Betula, 48
Bharal, **29**, 140
Bialowieza Forest, 47, 64
Bighorn, *see* Sheep, bighorn
Bilberries, 69
Birch, 48, 55, 88
Birds, 34, 70, 88, 110, 123–126, 157, 158, 160, 171, 174, 182
See also names of birds
Birds of paradise, 144, **146–147**
Bismarck Range, **146–147**
Bison, 28, 63
Bison bison, 63
Bison bonasus, 47, 58
Bistort, 30
Bitterbrush, 168
Black Forest Mountains, 18
Blarina brevicauda, 153
Blue Ridge Mountains, 152, 157
Boar, wild, 65, 66, 101, 158
Boas, dwarf, 149, 172
Bobcats, 158
Bohemian Forest, 40
Bongo, **29**, 108, **109**
Bos grunniens mutus, 140, **140**, **141**, 142
Brachypteryx stellatus, 126
Braided stream, 195
Bramblings, **90–91**
Brandenberg Mountains, **100**
Brassolidae, 175
Brazilian Highlands, 214
Brocket, red, **29**
Brooks Range, **148**, 165, 168
Budorcas taxicolor, 143
Buffalo, 105–106, 115
Bufo hoechstii, **100**
Bullfinch, 88
Bunchgrass, 182
Bunting, indigo, 161
Bushbuck, **29**, **104**, 106
Buteo jamaicensis, 157
Buteo lineatus, 157
Buteo platypterus, 157
Buteo rufofuscus, 101
Butterflies, **175**, **176–177**
Butterflies, monarch, 175
Butterflies, owl, **175**
Buzzards, augur, 101
Bycanistes brevis, 106

Cactaceae, 149
Cacti, 149
Cacti, saguaro, 26

224

Caldera, 195
Caledonian Range, 20
Caltha palustris, **129**
Camels, wild, 121
Camelus bactrianus, 121
Cameroon, Mt., 214
Canadian shield, 19, 153
Canis latrans, 153
Canis lupus, 65, 133, 153
Canis simensis, 101
Cantabrian Mountains, 40, 67
Canyon, 195, **195**
Capercaillie, **70**
Capra aegagrus, 138
Capra falconeri, 30, 138, **139**
Capra ibex, 34, **36**, 41, **44**, 101
Capra pyrenaica, 47
Capreolus capreolus, 58
Capricornius sumatraensis, 142, 143, **143**
Carabidae, 93
Cardinalis cardinalis, 158, **158**
Cardinals, 158, **158**
Caribou, 24–25, 149, 170
Carpathians, 40, 47, 55, 58, 65, 66, 67, 70, 75
Carstensz, Mt., 143
Cascades, 20, **21**, 148, 165
Castor canadensis, 158
Catasetum macrocarpum, **176–177**
Cathartes aura, 157
Catharus fuscescens, 161
Catharus minimus, 153
Caucasus, 40, 41, 47, 53, 54–55, 58, 63, 65, 69, 70, 76, 81, 87, 88, 214
Cavia porcellus, 149
Cayambe Volcano, 178
Central Massif, **40**
Cephalophus
Cephalophus callipygus, 114
Cephalophus natalensis, 114
Cereopithecus neglectus, **101**
Cereus giganteus, 26
Cervus elaphus, 34, 53, 58, 59, 145, 168, 170, **170**, 186
Chaffinches, 88
Chamois, 28, 30, 36, 49, 55, **56–57**, 58, 64, 75, 76, 143
Chapparal, 26
Cherskogo Range, **118**
Chiang Pai Range, 135
Chickadee, black-capped, 161
Chickadee, Carolina, 161
Chinch bugs, **52**
Chinchilla, 186, **186**
Chinchilla laniger, 186, **186**
Chioglossa lusitanica, 48

Chiru, 140
Cholla, 168
Chough, alpine, 17, 88, **89**
Chough, European, 101
Chozchoris, 186
Chrysolophus amherstiae, 126, **126**
Chugach Range, 148, 150–**151**
Cinder cone, 195
Cinnyris mediocris, **106**
Cinnyris venustus, **106**
Cirques, 152, 195, **195**
Cirsium spinosissinum, **51**
Citellus, 168
Cladonia, 149
Clytra laeviuscula, **52**
Coast Ranges, 148, 165, 170
Cock-of-the-rock, 175
Colobus polyomos, 106, **107**
Coluber viridiflavus, **94–95**
Columbia Glacier, **150**
Columnar jointing, 196, **196**
Communism, Mt., 121
Composite volcano, 196
Condor, Andean, 182, 183, 186
Condor, California, 171
Conglomerate, 196, **196**
Continental drift, **16**, 17, 18, 18, 196
Convection currents, **16**, 17, 197
Cook, Mt., 144
Copperhead, 161
Cordillera, 197
Core, Earth's, **16**, 17, 197
Coregonus, 44
Cornice, 197
Corvus brachyrhynchos, 161
Corvus corax, 81, **81**, 161
Cotopaxi, 178
Cougar, 152, 153, **154–155**, 168
Coyotes, 28, 153, 168, 171
Crater, 197, **197–198**
Creosote bushes, 168
Crevasse, 198, **198**
Crossbills, 88, 126, 161
Crossoptilon auritus, **123**
Crossoptilon mantchuricum, 126
Crotalus horridus, **161**
Crowfoot, Alpine, **28**
Crows, 161
Crust, Earth's, 16, 17, 18, 18, 19, **19**, 97, **210**
Ctenomys, 186
Cyanthea, **29**
Cyathea, 105
Cynomys ludovicianus, 170
Cyrilla racemiflora, 25

Dacryodes excelsa, 23
Dama dama, 186
Danaus plexippus, 175
Deer, 66, 69, 70
Deer, Andean, *see* Guemal
Deer, axis, 186
Deer, black-tailed, **38–39**
Deer, fallow, 186
Deer, mule, 28
Deer, musk, 29, **138**
Deer, red, 53, 58, **59**, **60–61**, 65, 67
Deer, 145, 170, **170**, 186
Deer, roe, 58, 65, 67, 70
Deer, white-tailed, 152, 153
Dendrocopos medius, 88
Dendrohyrax, 114
Dendroica cononata, 170
Desmognathus fuscus fuscus, **162**
Desmognathus wrighti, **162**
Dhaulagiri, Mt., 119
Diastrophism, 198
Diceros bicornis, 115
Diemictylus viridescens viridescens, **162**
Dike, *see* Dyke
Dinaric Alps, **40**
Divide, 198
Dolomys bogdanovi, 81
Dome mountain, 198
Dorcopsulus macleani, 143
Dragonfly, **128**
Drakensberg Mountains, **96**, 97, 214
Duiker, Peter's, 114
Duiker, red, 114
Dyke (Dike), **198**, 199

Eagles, 157
Eagles, bald, **157**
Eagles, Bonelli's, 81
Eagles, golden, 28, 47, **74**, **75**, 87, 157, 168
Earth, cross-sectional diagram of, **16**
Earthquakes, **18**, 20
East African Plateau, 214
Edelweiss, 30, **31**, 53, 128
Eft, red, **162**
Eifel, **40**
Elands, 115
Elaphe longissima, 92, 93
Elburz Mountains, 27
Elephants, 29, 34, 37, 106, 108, 115, **116–117**
Elfin woodland, 27, 172
Elgon, Mt., 97, 114
Elk, 48, 70
Elk, American, *see* Wapiti
Elm, 55
Empidonax virescens, 161

Endemism, 36
Epeirogeny, 199
Epimachus fastuosus, 144
Epiphytes, 27
Equus caballus, 49
Erethizon dorsatum, 152
Erica, 105, **115**
Erosion, 199, **199**
Erratic, **198**, 199
Escarpment, 199, **199**
Ethiopian Highlands, 101, 105
Etna, Mt., **22**
Europe, 40–95
Eurycea lucifuga, **162**
Euterpe globosa 26
Everest, Mt., 34, 119, 128, 214
Evergreens, 26
Everlasting, 114
Exfoliation, 199, **199**

Fagaceae, 66
Fagus grandifolia, 152
Falco peregrinus, 81, 157
Falco sparverius, **156**
Falcon, peregrine, 81
Fault, 200, **200**
Fault-block mountains, 200, **200**
Felis concolor, **152**, 153
Felis lynx, 70, **71**
Felis serval, 115
Felis sylvestris, 70, **72–73**
Ferns, tree, 105
Finches, 88
Finches, snow, 88
Finches, Tibetan snow, 128
Fir, 26, 149, 152
Fir, balsam, 152
Fir, Douglas, 28, 170
Fir, Fraser, 158
Fjord, 200, **200**
Fjordland (New Zealand), 144–145
Flamingoes, **29**, 149, 182, **184–185**
Flamingoes, Andean, 182
Flamingoes, James's, 182
Flycatcher, Acadian, 161
Fold, 200, **201**
Fold mountains, 201, **201**
Foliation, 201
Fox, red, 28, 75
Francolin, 110
Francolin, montagne, 115
Francolinus psilolaemus, 115
Francolinus squamatus, 110
Franconia Range, 149
Fringilla coelebs, 88
Fringilla montifringilla, 90–91

Frogs, arrow-poison, **172**, **173**
Frogs, tailed, 171–172
Fujisan, 19
Fumarole, 201, **201**
Fungi, 143

Gasherbrum, Mt., 119
Gazelle, Tibetan, 140
Gentian, Koch's stemless, **51**
Gentiana kochiana, **51**
Geosyncline, 201
Geranium, tree, 36
Geranium donniannum, **129**
Ghats, **118**
Gila monster, 172
Ginseng, 121
Glaciers, 152, 201–202, **202**
Glaciers, mountain valleys and, 54
Gladiolus, **105**
Glaucomys sabrinus, 158
Glaucomys volans, 158, **159**
Gneiss, 202, **202**
Goats, markhor, 30, 138, **139**, 140
Goats, mountain, 28, 34, 36, **36**, **37**, 64
Goats, Rocky Mountain, 17, 143, 149
Goats, wild, 123, 138
Godwin Austen, *see* K2
Goral, 29, 143
Gorge, 202, **202**
Gorilla gorilla, 105, 108, **110**, **111, 112–113**
Gorillas, 105, 108, 110, **110**, **111, 112–113**
Goose, bar-headed, 128
Goose, graylag, **130–131**
Graben, 202
Gramineae, 26
Grampians, 40, 58, 70, 75
Gran Paradiso, 44
Gran Sasso d'Italia, 66
Grand Canyon, 19
Grandala, 126
Grandala coelicolor, 126
Granite, 202, **203**
Graphosoma italicum, **52**
Grasshopper, short-horned, **128**
Great Dividing Range, **120**
Great Rift Valley (Africa), 18, 97, **98–99**, 105, 114
Great Smokies, *see* Smokies
Great Smokey Mountains National Park, 161
Grouse, black, 88
Grouse, red, **86**, 87
Grouse, willow, *see* Grouse, red

Guadarama, 40
Guanacos, 178, 182
Guemal, 186
Green Mountains, 149, 153
Guiana Highlands, 36, 178, 214
Guinea pigs, 149
Gymnogyps californianus, 171
Gypaetus barbatus, **74, 76**
Gyps fulvus, 76, **78–79**
Gyrinophilus porphyriticus, **162**

Haleakala, 36
Haliaeetus leucocephalus 157
Hammersley Range, **120**
Hares, 75, 70, 75
Hares, mountain (blue), 87
Harz Mountains, 40, 48
Hawaiian Islands, 20
Hawk Mountain, 157
Hawks, 157
Hawks, broad-winged, 157
Hawks, cooper's, 157
Hawks, red-shouldered, 157
Hawks, red-tailed, 157
Hawks, sharp-shinned, 157
Heath, 29, 105, 114–115, **115**, **116–117**, 158
Helarctos malayanus, 178
Heloderma suspectum, 72
Hemitragus hylocrius, 140
Hemitragus jayakari, 140
Hemitragus jemlahicus, **139**, 140
Heterohyrax brucei, 114
Hieraaetus fasciatus, 81
High Places of the World, 214–221
Highlands, 21
Himalayas, 18, **18**, 19, 20, 21, 29, 30, **118**, 119, 121, 126, 128, 133, 135, 143, 214
Hindu Kush, 119, 133, 135
Hindu Raj, 119
Hippocamelus antisiensis, 186
Hood, Mt., 165
Horn, 202
Hornbills, 106
Horns, 222–223
Horse, 49
Horst, 202
Hummingbirds, 175
Hypericum, 114
Hyraxes, 114, **114**, 115

Ibex, 28, 34, **36**, 41, 44, **44, 45**, 47, 49, 53, 55, 64, 75, 93, 101, 138
Ibex, Siberian, 138

Ice Ages, Pleistocene, 47, **47**
Icebergs, **150–151**
Ice-fall, 202, **203**
Igneous rock, 203
Ithaginis cruentus, 123, **124–125**

Jaguars, 149
Java, 22
Jebel Akhdar Range, **118,** 140
Joint, 203, **203**
Jötun-Fjeld, 22
Junco hyemalis, 153, 161
Juncos, 153, 161
Juniper, 26, 108
Juniperus, 108
Jura Mountains, 18, 23, 40

K2, 119
Kakapo, 145
Kamchatka, 22
Kanchenjunga, Mt., 119
Karakoram Range, **118,** 119, 135
Karst, 203, **203**
Katahdin, Mt., 152
Kea, 145, **145**
Kenya, Mt., 36, 105, 108, 115
Kestrels, American, **156**
Khap Yai National Park, 135
Kilimanjaro, 22, 97, 105, 106, 115
Klyuchevskaya Sopka, 121
Koklass, Nepal, 126
Kosciusko, Mt., 144, 215
Krakatoa, 19
Krummholz, 27
Kudu, 101
Kun Lun Mountains, 118, 121

Lacerta viridis, **46**
Lactuca alpina, 69
Lagidium pervanum, 186
Lagopus lagopus, 87
Lagopus mutus, 87, **87**
Laguna Colorada, 182
Lama glama, 36, **37, 180–181,** 182
Lama guanicoe, 178
Lama paca, 182
Lammergeier, 74, 76
Langurs, **121**
Lanius excubitor, 88
Larch, 66
Larix europaea, 66
Larrea divaricata, 168
Lava, 19, **19,** 97, 203, **210**
Leaf, transverse section of a, 20
Lebanon Mountains, **22, 118**

Lemmings, 82
Lemmus lemmus, 82
Leontopdium, 128
Leontopdium alpinum, 30, **31**
Leopards, **29,** 47, 55, 69–70, 105, 108, 135
Leopards, snow, 134, 135
Lepus capensis, 70
Lepus timidus, 70, 87
Lettuce, alpine, 69
Lice, 143
Lichens, 30, 58, 82, 115, 128, 143, 186
Lilium bulbiferum, 51
Lilium grayi, 158
Lily, Gray's, 158
Lily, orange, **51**
Lily, Turk's Cap, 28
Limestone, 203
Liverworts, 143
Livingstone Mountains, 97
Lizards, 172
Lizards, green, **46**
Lizards, swift (fence), 161
Llamas, 36, **37, 180–181,** 182
Lobelia, **29,** 105, 115
Loiseleuria procumbens, 66
Lophophorus, **122,** 123
Lophura nycthemera, 122
Louseworts, 138
Loxia curvirostra, 88, 126, 161
Loxodonta africana, 34, 106, **116–117**
Lucerne, Lake, **42–43**
Luquillo Mountains (Puerto Rico), 23, 27, 34
Lygaeus sexalilis, 52
Lygosoma laterale, 161
Lynx, 47, 70, 71, 152
Lynx, pardel, 53, **70,** 71
Lynx lynx canadensis, 152
Lynx pardellus, 53, 70, 71
Lynx rufus, 158
Lyrurus mlokosiewzzi, 88

Macdonnell Ranges, 120
Mackenzie Mountains, 148
Madagascar Plateau, **96**
Magma, 18, 203
Makalu, Mt., 119
Makran Range, 133
Mantle, Earth's, **16,** 17, 203
Maples 55, 152
Maquis, 26, 37, 53, 64
Marianas, 20
Maritime Alps, 76
Markhor, 30, 138, **139,** 140
Marmota calagata, 149
Marmota marmota, 49, 82, 84–85
Marmots, 75

Marmots, alpine, 49, 82, 84–85, 97
Marmots, hoary, 28, 149
Marten, pine, **83**
Martes martes, 82
Masico, 175
Massif, 204
Matterhorn, 55
Mau Escarpment, 97, 114
Mauna Kea, 20
Mauna Loa, 19, 20
McKinley, Mt., 165, 214
Mellivora capensis, 101
Melospiza melodia, 161
Meru, Mt., 97
Mesa, 204, **204**
Metamorphic rock, 204
Mice, *see* Mouse
Microtus, 82
Milkwort, shrubby, **50**
Mitchell, Mt., 157
Mites, 143
Moake Range, **120**
Monals, **122,** 123
Monkeys, colobus, 29, 106, 107, 108
Monkeys, De Brazza's, **101**
Monticola saxatilis, 88
Montifringilla adamsii, 128
Montifringilla nivalis, 88
Moose, American, *see* Elk
Moraine, 204, **204**
Moschus moschiferus, **138**
Moss, 27, 30, 58, 82, 115, **115,** 128, 143, 170, 186
Moss, reindeer, 149
Moth, bag-worm, 93
Mouflon, 34, 53, 62, **63–64**
Moulin, 204
Moreau, R. E., 110
Mountain building processes, 17–18, 97, **210–213**
Mountain lion, *see* Cougar
Mountains, 204, **205**
Mountains of the Moon, 214
Mouse, 81
Mouse, Prometheus, *see* Voles, burrowing
Mouse, red-backed, 153
Musgrave Ranges, **120**
Muskrats, 158
Mustela nivalis, **83**
Mustela putorius, 64
Mustelidae, 75
Myzornis, fire-tailed, 126
Myzornis pyrrhoura, 126

Naemorhedus goral, 143
Naga Hills, 119
Nan Shan, **118,** 121
Nappe, 204

Nassau Mountains (West Irian), 143
Natrix maurau, 93
Nectarinia, 105
Nestor notabilis, 145, **145**
Neve, 205
New Guinea, **120**, 143–144
New Zealand, **120**, 144–145
New Zealand Alps, **120**, 144, 145
Ngorongoro Crater, 115
Nothofagus, 149, 186
Notornis mantelli, 144
Nuee Ardente, 205, **205**
Nutcracker, Clark's 28
Nyala, mountain, 101

Oak, 26, 55
Oak, evergreen, 30
Obsidian, 205
Ochotona, **32**
Octodon degus, 186
Odocoileus
Odocoileus hemionus, **38–39**
Odocoileus virginianus, 152, 153
Olympic Mountains, 34, **148**, 170
Ondatra zibethica, 158
Oppossum, South American mouse, **29**
Opuntia bigelovii, 168
Opuntia ramossima, 168
Orange Mountains, 143
Orb, weaver, **128**
Orbell, G. B., 144
Orchard grass, **28**
Orchids, 149, 172, 175, **176–177**
Orchids, disa, 105
Orchis, 105, 149
Ore Mountains, **40**
Oreamnos americanus, 17, 36, 37, 143, 149
Orioles, black-headed, 115
Orioles, black-winged, 110
Orioles, green-headed, 110
Oriolus chlorocephalus, 110
Oriolus larvatus, 115
Oriolus nigripennis, 110
Orogeny, 205
Oryzopsis hymenoides, 168
Ospreys, 157
Outcrop, 205, **205**
Ovis ammon, 34, 53, **62**, **63**, 135, 138
Ovis canadensis, 34, **34**, 165
Owen Stanley Range, **120**
Owls, 26
Owls, saw-whet, 161
Oxlip, **50**

Palmae, 149
Palms, 149
Palms, sierra, 26, 27
Palo colorados, 26
Pamirs, **22**, **118**, 121, 135, 138
Panax, 121
Pandas, 121, 128, **132**, 133
Pandas, lesser, 128, **133**
Panthera onca, 149
Panthera pardus, 69, 105, 135
Panthera tigris, **123**, 133, 136–137
Panthera uncia, see Uncia uncia
Pantholops hodgsoni, 140
Papaver, 149
Parrot, Puerto Rican, 34
Partridge, chukar, **127**
Partridge, snow, 87
Parulidae, 161
Parus atricapillus, 161
Parus bicolor, **158**, 161
Parus carolinensis, 161
Pasqueflower, yellow alpine, 51
Passerina cyanea, 161
Patagonia, **187**, 189
Peregrine, *see* Falcon, peregrine
Peter and Paul Game Park, 44
Petrogale, 144, **144**
Phacochoerus aethiopicus, 106
Pharomachrus mocino, **174**, 175
Pheasants, 123–126
Pheasants, blood, 123, **124–125**
Pheasants, blue-eared, **123**
Pheasants, brown-eared, 126
Pheasants, silver, **122**
Phoebe, 161
Phoenicoparrus andinus, 182
Phoenicoparrus jamesi, 182
Phoenicopteridae, 149, **183**, **184–185**
Phylloscopus trochilus, 88
Pic-Orizaba, **22**
Picea abies, **66**
Picea mariana, 27, 152
Picea rubens, 152
Picea sitchensis, 149, 170
Piedmont glacier, 205
Pikas, 32
Pindus Mountains, **40**, 67, 75
Pine, 26
Pine, Lodgepole, **28**
Pine, ponderosa, 26, 28
Pine, stone, 66
Pinus monophylla, 168

Pinyon, 26, 168
Plate Tectonics, 206
Plateau, 205
Plethodon jordani, **160**, 161
Plethodon wehrlei dixi, 161
Plethodon yonahlossee, **162**
Plovers, 128
Pobesa Peak, 121
Podocarpaceae, 149
Podocarps, 149
Podocarpus, 108
Polygala chamaebuxus, **50**
Polygonum viviparum, 30
Polypepis trees, **29**
Pontic Mountains, 133
Poppies, arctic, 149
Populus, 49
Porcupine, 152
Porona grandiflora, **129**
Prairie dog, black-tailed, 170
Precambrian rock, 19–20
Presbytis entellus, **121**
Presidential Range, 27, 30, 149, 152–153
Primroses, 54
Primula, 54
Primula elatior, 50
Procapra picticaudata, 140
Procavia johnstoni, 114, **114**
Procaviidae, 114
Procyon lotor, 158
Prometheomys schaposchnikowi, 81
Pronghorn, 28
Protozoa, 143
Prunella collaris, 88, 126
Prunella himalayana, 126
Pseudois nayaur, 140
Pseudotrian ruber schencki, **162**
Pseudotsuga menziesii, 170
Ptarmigan, 28, 34
Ptarmigan, rock, 86, 87, **87**
Ptarmigan, willow, *see* Grouse, red
Pterocnemia pennata, **188**, 189, **189**
Pucrasia macrolopha nepalensis, 126
Pudu pudu, **187**
Pulsatilla alpina apiifolia, **51**
Pumice, 206
Puna, 182
Purshia tridentata, 168
Pyrenees, 34, 40, 54, 55, 67, 70, 75, 76, 87, 88
Pyrrhocorax graculus, 17, 88, **89**
Pyrrhocorax pyrrhocorax, 101
Pyrrhula pyrrhula, 88

Quercus ilex, 30
Quetzal, **174**, 175

Raccoons, 158
Rainier Mt., 165
Rakaposhi Mt., 119
Rangifer tarandus, 24–25,
149, 170
Rat, chinchilla, 186
Rat, rock, 186
Ratel, 101
Rattlesnake, timber, 161
Ravens, 81, **81**, 161
Redwood, 170–171
Relief, 206
Rhea, 189
Rhea, Darwin's, 188, 189,
189
Rhinoceros, black, 115
Rhododendron, 29, 31, 54, 66,
121, **121**, 123, 126, 128, 132,
158, 161
Rhododendron, 54
Rhododendron arboreum,
121
Rhododendron calendula-
ceum, 158
Rhododendron carolinia-
num, 158
Rhododendron ferrugineum,
31
Rhododendron kotschyi, 66
Rhodope, **40**
Rhone Glacier, 93
Ricegrass, 168
Rif, **96**, 101
Rift, 206, **206**
"Ring of fire," 20
Roan Mountain, 158
Roches moutonnees, 206 **206**
Rockies, 18, 23, 26, 28, 30,
148, 165, 214
Rotifers, 143
Rupicapra rupicapra, 30, 49,
55, **56–57**, 58, 143
Rupicola peruviana, 175
Ruwenzori Range, **96**, 97,
105, 114, 115, 214

Saddle, 207
Saguaro cacti, 26
St. Elias Range, **148**
St.-John's-wort, 114
Salamanders, 48, 101, **160**,
161, **162–163**, 165
Salamanders, alpine, 48
Salamanders, black-chinned
red, **162**
Salamanders, cave, **162**
Salamanders, fire, 48, **48**
Salamanders, gold-striped,
48

Salamanders, green, **160**
Salamanders, marbled, **162**
Salamanders, Metcalf's, 161
Salamanders, northern
dusky, **162**
Salamanders, pygmy, 162
Salamanders, red-cheeked,
160, 161
Salamanders, red-legged,
160, 161
Salamanders, Roanoke, 161
Salamanders, spotted, **162**
Salamanders, spring, **162**
Salamanders, Yonahlossee,
162
Salamandra atra, 48
Salamandra salamandra, 48,
48
Salmo agua-bonita, 172
Salmo trutta, 44
Salvelinus alpinus, 44
Sandpipers, 128
San Francisco Mountain, 149
Sarek Mountains, **22**
Sassurea, 121
Saxifraga, 168
Saxifrage, 168
Sayan Mountains, **22**, **118**,
135
Sayornis phoebe, 161
Scales, 30, 31
Scandinavian Range, 67, 70
Scandinavian upland, **40**
Sceloporus undulatus, 161
Schist, 19–20, 207
Scottish Highlands, 20, 58,
70, 87
Sea-floor spreading, 207
Sedges, 168, 182
Sedimentary rock, 206, 207
Sedum, 138
Sedum, 138
Selenarctos thibetanus, 133,
133
Semien Mountains, **96**, 97,
101
Sempervirens, 170
Senecio, 115
Senecios, **29**, 115
Sequoia gigantea, 149, 170
Sequoias, giant, 149, 170–
171
Serows, 29, 142, 143, **143**
Serval, 29, 115
Shale, 207
Sheep, Barbary, **102–103**
Sheep, bighorn, 28, 34, **34**,
164, 165, **166–167**, 168
Sheep, blue, see Bharal
Sheep, Larestan, 138
Sheep, Marco Polo, 135
Sheep, thinhorn (Dall), 165

Sheep, urial, 138
Sheep, wild, 34, 123, 135, 138
Shickshock, **148**
Shield volcano, 207
Shiga, Mt., **142**
Ship Rock (New Mexico), 19
Shortwing, Gould's, 126
Shrews, 81, 153
Shrike, great gray, 88
Sickle-bill, black, 144
Sierra de Juárez, 172
Sierra Madre del Sur, **148**
Sierra Madres, **148**
Sierra Morena Range, 53, 70
Sierra Nevada (Europe),
40, 53, **63**
Sierra Nevada (North Amer-
ica), 18, 148, 165, 168, 170,
172
Sikhote-Alin Range, **118**,
121, **135**
Sill, 207, **207**
Skinks, brown, 161
Skipper, **176–177**
Smokies, **148**, 152, 157–158,
161
Snakes 161
Snakes, Aesculapian, **92**, **93**
Snakes, viperine, 93
Snakes, whip, **94–95**
Snow line, 208
Snowy Mountains, **120**
Solenobia triquetrella, 93
Sorex cinereus, 153
South Pacific, 143–145
Southern Alps (New Zea-
land), **22**
Sparrows, song, 161
Sparrows, white-throated,
153
Speargrass, 168
Spiders, 34
Springtails, 34
Spruce, 26, 28, 55, 149, 158,
161
Spruce, black, 27, 152
Spruce, Norway, 66
Spruce, red, 152, 158
Spruce, Sitka, 149, 170
Squirrels, flying, 158
Squirrels, ground, 168
Stanovoi Range, **118**
Stipa speciosa, 168
Stoats, 64
Strigops habroptilus, 145
Subduction Zone, 208
Sudeten, **40**
Sudetes, **22**
Sunbirds, **29**, 105, **106**
Surdisorex norae, 36
Sus scrofa, 65, 101, 158
Sweet gom trees, 47

Sympetrum, 128
Syncerus caffer, 105
Syncline, 208

Table Mountain, **105**
Tabonuco trees, 23
Tahr, **29**
Tahr, Arabian, 140
Tahr, Himalayan, **139**, 140
Tahr, Nilgiri, 140
Tahsueh Shan, 119, 121
Takahe, 144–145
Takin, 143
Talamanca Range, **148**
Talus, 208, **208**
Tapirs, mountain, 178
Tapirus pinchaque, 178
Tarn, 208, **208**
Tatra, 22, **40**, 55, 66, 70, 83
Taurotragus eurycerus, 108, **109**
Taurotragus oryx, 101
Taurus Mountains, 20, **118**
Terminal moraine, 208, **208**
Tetraogallus caucasicus, 87
Tetrao urugallus, 70
Theropithecus gelada, **101**
Thistle, spiny, **51**
Thrush, gray-cheeked, 153
Thrush, rock, 88
Thryothorus ludovicianus, 161
Tibesti, **96**
Tibetan Plateau, 121, 126, 128, 140
Tichodroma muraria, 88
Tien Shan, **118**, 121, 135, 138
Tierra del Fuego, **22**, 23, 178, 189
Tigers, 123, 135, **136–137**
Tigers, Caspian, 135
Tigers, Siberian, 133, 135
Timberlines, **22**, 23, 27, 34, 152, 208
Tirich Mir, 119
Tit, crested, **28**
Titmouse, tufted, **158**, 161
Toads, **100**
Tragelaphus buxtoni, 101
Tragelaphus scriptus, **104**, 106
Tragelaphus strepsiceros, 101
Tragopan, satyr, **127**
Tragopan, Temmink's, 123
Tragopan satyra, **127**
Tragopan temmincki, 123
Transylvanian Alps, 40, 66, 70
Trebevic, Mt., 81
Tremarctos ornatus, 177
Trillium, red, 158

Trillium erectum, 158
Trout, golden, 172
Trout, lake, 44
Tsinling Mountains, **118**
Tucu-tuco, 186, 189
Tuckerman's Ravine, 152
Tulip trees, 47
Tundra, 23, 27, 47, 65, 153, 158
Turtle bugs, **52**

Uintas, **148**
Ulugh Muztagh, 121
Uncia uncia, **134**, 135
Ungaliophis continentalis, 149, 172
Ural Mountains, **22**, **40**, 47, 65, 133
Ursus americanus, 158
Ursus arctos, 67, **68**, 133, **168**
Ursus arctos horribilis, 69, 149, 168, **168**, **169**
U-shaped valley, 209, **209**

Vaccinium, 54
Vaccinium myrtillus, 69
Variegata orbea, **100**
Vascacha, **29**
Veery, 161
Verkhoyansk Range, **118**
Vesuvius, Mt., 19
Vetches, 138
Vicia, 138
Vicugna vicugna, 178, **182**
Vicuña, **29**, 178, 182, **182**
Viola, 158
Violets, 158
Vipera aspis, 92
Viperine snake, 93
Vireonidae, 161
Vireos, 161
Virunga Volcanoes, 110
Viscacha, mountain, 186, **186**
Volcanic plug, 208, 209
Volcanoes, 18–19, 20, **20**, **21**, 26, 97, 110, 121, 165, 178, 209, **210**, 214
Volcanoes, formation of, 18–19, **210**
Voles, burrowing, 81
Voles, snow, 81–82
Vosges Mountains, 18
Vulcanism, 18, **18**, 19, 97
Vulpes vulpes, 75
Vultur gryphus, 182, **183**
Vultures, 76, 81, 157, 171
Vultures, bearded, *see* Lammergeier
Vultures, black, 76, 77
Vultures, griffon, 76, 78–79, 80
Vultures, turkey, **157**

Wall creeper, 88
Wallabies, forest mountain, 143
Wallabies, rock, 144, **144**
Wapiti, 34, 168, 170, **170**
Wapiti, Roosevelt, 170
Warblers, 161
Warblers, Canada, 161
Warblers, willow, 88
Warblers, yellow-rumped, 170
Wart hogs, 106, 115
Wasatch Mountains, **148**
Washburn, Mt., 168
Washington, Mt., **22**, 30, 157
Water table, 209
Weasels, 75, 83
Weathering, 209, **209**
Weevils, 143–144
White Mountains, 27, **148**, 149, 152
White-eyes, 110
Whitefish, 44
Wildcat, 55, 70, 72–73, 75
Wilhelmina, Mt., 143
Wisent, 47, 55, 58, 63, 64
Wolves, 29, 34, 47, 53, 55, 64, 65, **65**, 66, **66**, 67, 133, 153, 168
Wolves, Abyssinian, 101
Woodpeckers, 26
Woodpeckers, middle-spotted, 88
World Wildlife Fund, 67
Wrangell Mountains, **148**
Wren, Carolina, 161

Xenophon, 63

Yak, wild, 140, **140**
Yellowwood, 108

Zagros Mountains, 20, 121, 138
Zonation, 23
Zonotrichia albicollis, 153
Zosteropidae, 110

Photographic Credits

Numbers correspond to page numbers.

1 Breck P. Kent/Animals Animals; 2–3 François Gohier; 4–5 Jean-Paul Ferrero; 6–9 Stouffer Productions, Ltd./Animals Animals; 20 top Richard Rowan; bottom Carol Cawthra/Earth Scenes; 21 Harald Sund; 24–25 George Calef; 31 Jean-Paul Ferrero; 32–33 top Jeffrey Foott; bottom Rene P. Bille; 35 top A. J. Sutcliffe/Natural Science Photos; bottom Keith Gunnar/Tom Stack & Associates; 36 top Keith Gunnar/Bruce Coleman, Inc.; bottom Stephen J. Krasemann; 37 Sonja Bullaty and Angelo Lomeo; 38–39 William A. Bake; 42–43 Richard Rowan; 44 Hans D. Dossenbach; 45 top A. J. Sutcliffe/Natural Science Photos; bottom Jean-Paul Ferrero; 46 Hans D. Dossenbach; 48, 50–51 top left Rene P. Bille; top center M. P. L. Fogden/Ecology Pictures; top right, bottom left Sonja Bullaty and Angelo Lomeo; bottom center Jean-Paul Ferrero; bottom right Hans D. Dossenbach; bottom Thomas Vanderschmidt/Earth Scenes; 52 Walter Fendrich/Animals Animals; 55 James Lester/Photo Researchers, Inc.; 56–57 Hans D. Dossenbach; 59 Hans Reinhard/Bruce Coleman, Inc.; 60–61 Helmut Gritscher/Animals Animals; 62 Kenneth W. Fink/Bruce Coleman, Inc.; 63 Walter Fendrich; 65, 66 John Ebeling; 68 Thase Daniel; 70 top Roy A. Harris and K. R. Duff/J. Allan Cash, Ltd.; bottom A. Winspear Cundall/Natural Science Photos; 71 top Gunter Ziesler/Photo Researchers, Inc.; bottom Brian Hawkes/Photo Researchers, Inc.; 72 G. Kinns/Natural Science Photos; 73 top R. Balharry/Natural History Photographic Agency; bottom Phillip Wayre/Natural History Photographic Agency; 74 top Eric Hosking/Bruce Coleman, Inc.; bottom Tom Brakefield; 77 Eric Hosking; 78–79 Jean-Paul Ferrero; 80 top Rene P. Bille; bottom Jean-Paul Ferrero; 81 Spectrum Color Library; 82 Rene P. Bille; 83 top Hans Reinhard/Bruce Coleman, Inc.; bottom Roy A. Harris and K. R. Duff/J. Allan Cash Ltd.; 84–85 Thomas Vanderschmidt/Animals Animals; 86 top Jeffrey Foott bottom & 87 Stephen J. Krasemann; 88 top Jean-Paul Ferrero; bottom & 89 Rene P. Bille; 90–91 Arthur Christiansen; 92 top Rene P. Bille; bottom Hans D. Dossenbach; 93, 94–95 Rene P. Bille; 98–99, Carl Purcell/Photo Researchers, Inc.; 100 D. C. H. Plowes; 101 top Leslie H. Brown; bottom Thase Daniel; 102–103 top left John C. Stevenson/Animals Animals; top right Patricia Caulfield/Animals Animals; bottom Sonja Bullaty and Angelo Lomeo; 104 Peter Davey; 106 top Dale and Marion Zimmerman; bottom Hans D. Dossenbach; 107 Eric Hosking; 108–109 Alan Root; 110 Lee Lyon/Bruce Coleman, Inc.; 111 Deschryver/Tierbilder Okapia; 112–113 Alan Root/Tierbilder Okapia; 114, 115 top Hans D. Dossenbach; bottom Edward S. Ross; 116–117 Hans D. Dossenbach; 121 John Melville Bishop; 122 top Eric Hosking; bottom, 123 Kenneth W. Fink/Bruce Coleman, Inc.; 124–125 G. D. Plage/Bruce Coleman, Inc.; 126 Colin B. Frith/Bruce Coleman, Inc.; 127 top Belinda Wright; bottom Tom Brakefield/Bruce Coleman, Inc.; 128–129 John Melville Bishop; 130–131 Belinda Wright; 132 Jean-Paul Ferrero; 133 Thase Daniel; 134–135 Belinda Wright; 136 George B. Schaller/Bruce Coleman, Inc.; 137 Stanley Breeden; 139 top Robin Smith Photography, Ltd.; bottom George B. Schaller/Bruce Coleman, Inc.; 140 Fran Allan/Animals Animals; 141 George B. Schaller/Bruce Coleman, Inc.; 142–143 Tetsuo Gyoda; 144 top Jean-Paul Ferrero; bottom Francisco Erize/Bruce Coleman, Inc.; 145 Thase Daniel; 146–147 Erwin A. Bauer; 150–151 Harald Sund; 152 Ernest Wilkinson/Animals Animals; 154–155 W. T. Hall/Bruce Coleman, Inc.; 156 Tom Brakefield; 157 top Stouffer Productions, Ltd./Animals Animals; bottom, Jack L. Griffin/Animals Animals; 158 Zig Leszczynski/Animals Animals; 159 Stouffer Productions, Ltd./Animals Animals; 160 top, center Harry Ellis; bottom Jack Dermid; 162–163 bottom right Harry Ellis; all others Jack Dermid; 164 Charles G. Summers, Jr./Tom Stack & Associates; 166–167 Stephen J. Krasemann; 168–169 Stouffer Productions, Ltd./Animals Animals; 170 top Thase Daniel; center John Ebeling; bottom Stephen J. Krasemann; 172 Edward S. Ross; 173 top Edward S. Ross; bottom M. P. L. Fogden/Ecology Pictures; 174 top Robert L. Dunne/Bruce Coleman, Inc.; bottom Bruce Coleman, Inc.; 175, 176–177 Raymond A. Mendez/Animals Animals; 180–181, 182 William Franklin; 183 top François Gohier; bottom Tony Morrison; 184–185 Marion Morrison; 186 Tony Morrison; 187 François Gohier; 188–189, 190–191 Tony Morrison;